Training-Within-Industry Job Programs for Improved Construction Safety

The purpose of this book is to demonstrate how a Training-Within-Industry (TWI) Job-Program could reduce human factor-related harm in construction.

The construction industry has a significant impact on issues relating to the health, safety, and well-being (HSW) of people in the workforce. It is important to acknowledge that workers' behaviour influences the safety management system (SMS) of construction projects either negatively or positively and that it is important for a management team to identify relevant behaviours and take appropriate action to solve problems. In most cases, accidents happen because of the results of human failure in the form of errors, violations and system failures. Human failure causes accidents and site management needs to reduce hazards that might cause such errors, violations and system failures on worksites. The chapters in the book address factors causing human failure on construction sites, how to mitigate errors and violations through SMS and 'learning by doing' and improving practice of using safety instructors on sites. The book closes with insights from a TWI-informed human failure reduction framework.

This book provides valuable insights into safety management in a construction site context that can be applied to other areas. It is essential reading for safety managers, construction managers, researchers, and advanced students.

Routledge Research Collections for Construction in Developing Countries

Series Editors: Clinton Aigbavboa, Wellington Thwala, Chimay Anumba, and David Edwards

Training-Within-Industry Job Programs for Improved Construction Safety

Lesiba George Mollo, Fidelis Emuze, and John Smallwood

Routledge
Taylor & Francis Group

LONDON AND NEW YORK

First published 2024
by Routledge
4 Park Square, Milton Park, Abingdon, Oxon OX14 4RN

and by Routledge
605 Third Avenue, New York, NY 10158

Routledge is an imprint of the Taylor & Francis Group, an informa business

British Library Cataloguing-in-Publication Data
A catalogue record for this book is available from the British Library

ISBN: 978-1-032-42778-2 (hbk)
ISBN: 978-1-032-43015-7 (pbk)
ISBN: 978-1-003-36534-1 (ebk)

DOI: 10.1201/9781003365341

Typeset in Times New Roman
by Apex CoVantage, LLC

Contents

Figures

Tables

Preface

The need to blend the interaction between workers, management, equipment (including plants and tools), and environment increases the complexity of a typical construction project site. However, the complexity must be confronted as construction operations are notorious for having high levels of hazards and risks that may harm workers and the public. Statistics from organisations such as the International Labour Organisation (ILO), Health and Safety Executive, and the Occupational Health and Safety Administration show that accident prevention efforts cannot be slack on construction sites if people are to be shielded from harm. Training is one effort available to all construction sites in developed or developing countries, either in labour-intensive or automated construction. Training of people in construction (PiC) who lacks health and safety (H&S) skills, who are newly hired, who are transferred or seconded to another task, and who have not performed a task for a long time is vital to preventing accidents, diseases, injuries, and fatalities in construction operations. The training of PiC, which includes workers, supervisors, and managers, is a crucial element of all safety management systems (SMS). When the training is inadequate, wrong, or lacking, unsafe acts and conditions that precede accidents will become pervasive in construction projects. Where safe work performance skills are lacking or inadequate, training is appropriate. For this reason, and many more outlined in Chapters 1 to 6, this book has been put together to engender continuous H&S improvement in construction site operations by unpacking a model of training-within-industry (TWI). TWI is a methodology that trains everyone involved in production activities in open or enclosed spaces to reduce defects, rework, delays, and H&S issues. It is a lean approach that uses the continuous improvement principle to close any skill and knowledge gaps that may hamper workers, supervisors, and managers' safe and quality performance of tasks. As readers will see in this book, TWI has three basic components, which include job relations (JR), job instruction (JI), and job methods (JM). These components help create and sustain trust, cooperation, standardisation, and waste reduction (both physical and process waste) in an industrial engineering environment such as a construction site.

With a focus on construction site operations, the book outlines how TWI links up and better H&S, how to limit errors and violations, and how to halt human errors with SMSs, learning-by-doing, management control, and TWI. The data shared in the book illustrate how these concepts would continue to better occupational health and safety (OHS) in construction.

About the chapters in the book

Chapter 1 emphasises that OHS is a system designed to prevent accidents to create a safe and healthy workplace. This chapter presents hindrances to TWI use and ways to overcome them through learning by doing on construction sites. The chapter identified the human factors that slow down the use of learning by doing on construction sites. These factors include insufficient training budget, knowledge and skill transfer failure, lack of interest to work in the construction industry, social issues affecting workers, construction productivity, and lack of training facilities on construction sites. The chapter, therefore, argues that OHS in construction operations must improve by leveraging the ideas linked to learning by doing and TWI. Chapter 2 reminds readers of human factors responsible for safety failures on construction sites. These factors (whether small or big) are often responsible for accidents recorded in construction. The chapter underlines the causality around workers, management, and the project context. As a result, the chapter urges project teams to improve their capacity for safer construction operations.

Chapter 3 is about how to deploy SMSs that would prevent human failure. The chapter is premised on the notion that ineffective SMSs contribute to human failure causing accidents. The chapter shows that SMSs with the necessary features would reduce human failures on construction sites. The relevant features of SMSs are hazard identification, safety inspection, OHS training on the job, safe work practices, risk assessment plan, safety reporting system, toolbox talk, and OHS policy. In contrast, a poor SMS on construction projects makes human failures the norm on construction sites. Such a 'norm' must not be allowed to evolve either through safety violations or human error. And so, human error is addressed in Chapter 4. The chapter noted that a major causal factor of accidents is human error, a subject of several books by James Reason and Sidney Dekker. This chapter shows how learning-by-doing could minimise human errors on construction sites. The chapter confirms that human factors remain the main source of operation errors. It thus illustrated how learning-by-doing would convey job knowledge and abilities in training to remedy the situation. In particular, the chapter stated that H&S training is provided at the beginning of the project. By combining it with learning-by-doing, workers, supervisors, and managers will be able to learn about job requirements as they perform their jobs.

Chapter 5 turns to the role of supervision and management on construction sites as it concerns H&S. The chapter shows the impact of safety managers on construction projects OHS due to the reality that inadequate supervision and management can lead to accidents. Also, the chapter indicates that safety managers should monitor construction OHS better using a structured approach. The approach should help a safety manager know the regulations to implement SMSs. This would enable safety managers to effectively monitor the OHS to prevent accidents in construction. The closing chapter of the book presents a modified TWI model. Chapter 6 recognised that construction operation is notorious for recording a high number of accidents when compared with other industries because of human error. This chapter conceptual argues that integrating TWI and SMSs into a model will minimise human errors on construction sites. A survey questionnaire evaluated a conceptual model to realise this objective. The chapter shows that such a model would contribute to the development of new knowledge, specifically the ability of a safety manager to monitor and keep the worksite safe.

Acknowledgements

A book of this value will only be possible with the support of experts affiliated, either through membership or participation in annual conferences, with the CIB Working Commission on Health, Safety and Wellbeing (W099). Their peer review of each chapter contributed to the quality of the volume. We, therefore, appreciate Abimbola Windapo (University of Cape Town, Cape Town, South Africa), Alenka Temeljotov-Salaj (Norwegian University of Science and Technology, Trondheim, Norway), Dayana Bastos Costa (Federal University of Bahia – Salvador-Bahia-Brazil), Elaine Varela (Federal University of Bahia – Salvador-Bahia-Brazil), Emmanuel Bannor Boateng (University of Wollongong, NSW 2522, Australia), Khursheed Ahmed (Karakoram International University, Gilgit, Pakistan), Lim Shu Hui Michelle (National University of Singapore, Singapore), Michelle Turner (Royal Melbourne Institute of Technology, Melbourne, Australia), Patricia Omega Kukoyi (University of Lagos, Lagos, Nigeria), Roseneia Melo (Federal University of Bahia – Salvador-Bahia-Brazil), and Tarcisio Abreu Saurin (Federal University of Rio Grande do Sul – Port Alegre, Brazil), Alfredo Soeiro (University of Porto, Porto, Portugal), and. We thank Renée van der Merwe for the English language editing of the book. We also thank Martha Luke and Ed Needle of Taylor and Francis for their support.

Fidelis Emuze
Bloemfontein, South Africa
21 March 2023

Authors' biographies

Lesiba George Mollo is a Senior Lecturer in the Department of Built Environment, Faculty of Engineering, Built Environment, and Information Technology at the Central University of Technology, Free State (CUT), South Africa. He obtained a PhD in Construction Management and a master's degree (M-Tech) in Quantity Surveying from Nelson Mandela University (NMU). He has worked for numerous construction companies on a part-time basis. His research interests include construction health and safety, wearable sensing technologies, and building energy management.

Fidelis Emuze, PhD, is a Professor and Head of the Department of Built Environment at the Central University of Technology, Free State (CUT), South Africa. Lean construction, health, safety, and sustainability constitute the primary research interest of Dr. Emuze, who is a National Research Foundation (NRF) C-rated researcher that has published over 250 research outputs and received over 25 awards and recognitions. Dr. Emuze is the editor of Value and Waste in Lean Construction, Valuing People in Construction, and co-editor of Construction Health and Safety in Developing Countries. Dr. Emuze authored Construction Safety Pocketbook for South Africa in 2020. Dr. Emuze is the International Coordinator of CIB W123 – People in Construction Working Commission.

John Smallwood is the Professor of Construction Management in the Department of Construction Management, Nelson Mandela University, and the Principal, Construction Research Education and Training Enterprises (CREATE). Both his MSc and PhD (Construction Management) addressed construction health and safety (H&S). He has conducted extensive research and published in the areas of construction H&S, ergonomics, and occupational health (OH), but also in the areas of construction management education and training, environmental management, health and well-being, primary health promotion, quality management, risk management, and the practice of construction management.

Summaries of chapters

Chapter 1 Construction safety and training-within-industry

Occupational health and safety (OHS) is a system designed to lower the accident rate accidents and create a safe and healthy environment for workers in the workplace. This chapter presents the factors preventing workers from adopting training-within-industry (TWI) and improving OHS through learning by doing on construction sites. A case study research design was adopted to collect the data through semi-structured and focus group interview techniques. The findings identified the human factors that negatively affect implementing learning-by-doing on construction sites. The factors include insufficient training budget, knowledge and skill transfer failure, lack of interest to work in the construction industry, social issues affecting workers, construction productivity, and lack of training facilities on construction sites. Therefore, construction operations must improve OHS by adopting learning-by-doing. TWI promotes learning-by-doing and could improve construction workers' knowledge and skills transfer programs.

Chapter 2 Causes of human failure on construction sites

This chapter presents the factors causing human failure on construction sites. Human failure is responsible for most of the accidents recorded in the construction industry. A case study research approach was adopted to solve this reported human failure problem. The qualitative data were collected through semi-structured interviews, focus group interviews, and participant observations. The results show that the causes of human failure are workers, management, and construction project factors. The factors related to human failure include errors and violations. As a result, this chapter urges stakeholders to improve their capacity for enforcement and inspection in the construction industry.

Chapter 3 Human failure and safety management systems

This chapter presents how implementing the safety management system (SMS) would prevent human failure on construction sites. An ineffective SMS contributes to human failure, causing accidents that lead to construction loss. The human failure problems were identified using a mixed-methods research approach that integrated statistical and textual data. The results show that integrating SMS components would reduce human failure on construction sites. The SMS comprises the following components: hazard identification, safety inspection, OHS training on the job, safe work practices, a risk assessment plan, a safety reporting system, a toolbox talk, and an OHS policy. In addition, the integrated model to prevent human failure could be adopted to help improve the SMS on construction sites. Poor SMS on construction projects has allowed human failures to become the norm on the visited construction sites. Workers and management should work together to address the poor SMS implementation in the study to reduce human failure that results in accidents.

Chapter 4 Minimising human errors with learning-by-doing

Accidents may accompany the delivery of projects for several reasons. One causal factor of accidents is human error. This chapter presents how learning-by-doing could minimise human error on construction sites. A mixed-methods research was used to collect both qualitative and quantitative data. The chapter confirms that management and workforce-induced factors remain the main source of human error in construction operations. Learning-by-doing could promote job knowledge and abilities in an informal training program without referencing any theoretical concept or structured learning program. Safety training is provided at the beginning of the project, and by combining it with learning-by-doing, the workers can learn about the requirements as they go about their work. Therefore, authorities on a construction site should provide opportunities for hands-on training to those who work on projects to reduce human error.

Chapter 5 Managing health and safety on construction sites

This chapter presents the impact of safety managers on construction projects' occupational health and safety (OHS) because accidents can be caused by inadequate supervision and management. The data were collected in Bloemfontein, South Africa, utilising semi-structured and focus group interviews following the case study research approach. The results show that safety managers should monitor OHS on construction projects. To fulfil this function better, the chapter unpacks a framework for monitoring the safety managers' impact on construction projects. The framework stipulates that the safety manager must be familiar with OHS regulations to implement a safety management system. This would enable the safety manager to effectively monitor the OHS on construction projects to prevent accidents.

Chapter 6 Minimising human errors on construction sites

Construction is known for recording a high number of accidents when compared with other industries. A primary cause of accidents is human error. This chapter argues that integrating training-within-industry (TWI) and safety management systems (SMS) will minimise human error on construction sites. A survey questionnaire evaluated a conceptual model to realise this objective. The chapter shows that a conceptual model that integrated the TWI job program factors and the SMS components contributes to the development of new knowledge, specifically the abilities of a safety manager to monitor and keep the worksite safe.

1 Construction safety and training-within-industry

1.1 Introduction

Occupational health and safety (OHS) is described as a system designed to lower the incidence of accidents and set up a safe and healthy environment for workers in the workplace (Liu et al., 2020). In addition, OHS is described as a systematic, interdisciplinary process for detecting and evaluating potential workplace hazards that employees may encounter in the construction industry (Liu et al., 2021). The implementation of OHS on construction sites deals with worker promotion, protection, and the prevention of illnesses and accidents related to the workplace. However, construction continues to be known as an industry in which increasing numbers of workers suffer from accidents (Ajslev & Nimb, 2022). For example, the National Safety Council in the United States of America (USA) reports that 8,993 individuals died at construction sites between 2003 and 2011, the highest number of fatalities across all industries (Mahmoudi et al., 2014).

Additionally, the Health and Safety Executive (HSE) data from 2019 indicated that the construction industry was responsible for over 22% of all occupational fatalities in the United Kingdom (UK). In the UK, 2,420 more construction workers had non-fatal occupational injuries in 2018–2019 (Health and Safety Executive [HSE], 2019). Therefore, conditions in the USA are comparable to those in the UK. Additionally, it has been reported that in 2017, the construction industry in developing nations such as South Africa recorded 1.5 to 2.5 fatalities on average each week (Department of Labour, 2017). These alarming figures highlight the need for a concentrated, intentional effort to create, adopt, and assess creative solutions to this complex issue.

Construction orgaisation in both developing and developed countries have taken action to address the issue of poor OHS and reduce the high-accident rate in the construction industry (Awwad et al., 2016). Koskela (1992) noted that the high number of accidents recorded in the construction industry could be due to construction methods. Thus, it is critical to search for alternative methods of construction that would reduce the number of accidents. Therefore, to resolve this highlighted OHS issue in the construction industry, researchers such as Huntzinger (2016), and Graupp and Wrona (2010) recommended the implementation of the training-within-industry (TWI) method in the construction industry.

DOI: 10.1201/9781003365341-1

The TWI method is a leadership development method that equips team leaders and supervisors with the skills necessary to direct, instruct, and enhance the duties of their subordinates (Allen, 1919). In brief, the TWI method was designed in the 1940s by the National Défense Advisory Commission (NDAC) in the USA to increase manufacturing output and support the armed forces' war effort (Huntzinger, 2016). The TWI approach enabled Americans to boost industrial production by using workers without manufacturing experience to produce aeroplanes and weaponry (Latijnhouwers & Berendsen, 2014). The TWI intends to enable learning by doing, which means managing production-related issues under the supervision of a supervisor who has completed the necessary training (Huntzinger, 2016). As a result, a well-trained manager or supervisor should be equipped with the skills to analyse the tasks for the workers. To analyse the workers' tasks, a supervisor must be informed about every subject that the workers need to be taught (Allen, 1919).

This chapter presents factors that prevent workers from adopting TWI or improving OHS through learning by doing on construction sites. The idea that the TWI approach helped the USA increase manufacturing production and support the military forces motivated its implementation in this study. Therefore, adopting the TWI method could help address the construction industry's OHS problems. This chapter's subsequent sections are organised as follows: In Section 2, the background of the TWI method is explored. The research methods used for this study are described in Section 3, while the analysis and research findings are presented in Section 4. Finally, Section 5 provides conclusions and recommendations.

1.2   Training-within-industry (TWI) method

The development of the TWI method concentrated a significant focus on training the production or manufacturing leaders, masters, foremen, and experienced operators. At that time, the objective was to enhance the existing operational process, acquire in-depth expertise in employee training, and maintain a cooperative working environment between employers and employees in manufacturing industries (Misiurek & Misiurek, 2017). As stated in the previous section, the TWI method was developed to help wartime industrial sectors fulfil their personnel needs by training each employee in the industry to utilise their finest skills to the utmost extent feasible according to their unique capabilities; thereby, enabling production to keep up with the war demand (Huntzinger, 2016).

The TWI method equips production managers and senior managers with the skills they need to promote work standardisation, method improvement, and transformation (Latijnhouwers & Berendsen, 2014). Obtaining knowledge facilitates sustaining positive working relationships and enhances operational processes and employee capabilities (Misiurek & Misiurek, 2017). To maximise employee training through the TWI method, Huntzinger (2016) suggested that the following four rules be followed: standards must be created; best practices must be adopted in instructing; there should be continuous training; and efficient scheduling of training must be practised.

It should be mentioned that the TWI method's core idea is to encourage 'learning by doing', which refers to fixing production issues while being guided by an instructor or supervisor who is competent for the position (Huntzinger, 2016). Between 1940 and 1945, the TWI approach expanded across the US as a result of its success in the manufacturing industry. The TWI method was focused on improving employees' skills to perfect their professions rather than raising productivity (Misiurek & Misiurek, 2017). Additionally, the TWI was identified using the Five Needs of a Good Supervisor, which also includes knowledge of the work, knowledge of responsibility, skill in instructing, skill in improving methods, and skill in leading.

Evidence of the TWI method suggests that the industry's training practices are complex owing to two processes: training with absorption and training by intention. The TWI method can be broken down into these two categories (Allen, 1919). These two categories can be used to analyse the TWI approach (Allen, 1919). For example, employees may be trained through training with absorption; however, there needs to be an established strategy for doing so. New talents can be learned from someone who executes the same job duties. Practices that fall under the heading of 'training by intention', in which an instructor demonstrates to new employees how to carry out job-related duties, include an apprenticeship program (Allen, 1919).

According to Grip and Sauermann (2013), the development of the TWI was built on the Three J-Program (J-P), which consists of Job Instruction (JI), Job Methods (JM), and Job Relations (JR). The three J-Ps are described below.

1.2.1 Job instruction training (JIT)

The standardisations of work are one of the cornerstones of job instruction training (JIT). JI in industries aims to inform instructors or managers on how to produce skilled workers (Allen, 1919). According to Miron et al. (2016), 'standardisation of work' is a constituent of JI, which supports Huntzinger's (2016) claim that JI is related to the standardisation of work. As a result, the JI or standardisations of work enables an analysis of a procedure to define its guidelines (Feng & Ballard, 2012). In addition, the standardised work approach, which is based on recognised consistency in work operations, is the cornerstone of every continuous improvement initiative.

The impact of JI is described by Allen (1919) as a manager needing to be skilled in how to give instructions or be trained on how to follow instructions. Even when a supervisor is an expert in their position, it only sometimes follows that they can always instruct their employees on completing a task. This is particularly important in industries that require much labour (Allen, 1919). According to additional data provided by Dinero (2010), 'job instruction' is used because, if a learner has not learned the work, the instructor has not sufficiently taught them how to complete the task. Such a statement raises the issue of who is accountable for accidents on construction sites while supervisors are involved in the project.

JT in the industry is significant because training improves the workers' skills, thus making workers more productive and improving overall productivity. Also,

JT produces positive returns to the organisation, which could be explained by other factors, such as higher motivation or stronger loyalty of the workers to their employers (Grip & Sauermann, 2013). Therefore, it is one of the responsibilities of employers to provide training for workers to develop and equip them with the skills they require in the workplace (Graupp & Wrona, 2010).

1.2.2   Job method training (JMT)

The main objective of job method training (JMT) is to assist managers in producing superior products in an appropriate amount of time by making the best use of hired individuals, tools, and materials (Huntzinger, 2016). For example, managers are taught how to break down job procedures into their core activities during JMT, and every aspect of job activities is rigorously questioned to generate ideas for improvement (Kováčová, 2012). Additionally, the purpose of the job method is to stop managers from presenting management with incomplete or incorrect ideas (Huntzinger, 2016).

The job method is like JI in that it calls for employees to evaluate a task that managers have given them. Nevertheless, it varies from JI in that its primary objective is to improve how the task is carried out (Dinero, 2010). One benefit of the job method over JI is that the job method, according to Huntzinger (2006), involves reorganising and streamlining work processes to maximise production. It also involves training managers on organising jobs into their essential processes. According to Kováčová (2012), the job technique helps managers hold themselves accountable for 'kaizen', or continuous development. Such accountability encourages managers to enhance their work practices continually.

According to Kumar (2019), the concept of kaizen, meaning 'continuous improvement', was developed in Japan. It focuses on product improvement and is implemented to improve processes. Kaizen improves products through three parameters: quality, cost, and delivery (Georgise & Mindaye, 2020). Thus, the notion of the job method is connected to continuous improvement and significantly impacts lean practice, sometimes referred to as kaizen. Continuous improvements are, therefore, a component of lean tools used to reduce waste in organisations while generating greater value (Korb, 2016). It is critical to prevent accidents from happening to continue improving work practices. The idea of continuous improvement covers the concepts and techniques for controlling processes that reduce business waste (Miron et al., 2016). Koskela (1992) defines the term continuous improvement as a measuring and monitoring improvement system that uses standard procedures as hypotheses of best practice to be challenged constantly in better ways.

1.2.3   Job relation training (JRT)

Job relation training (JRT) aims to help managers improve their capacity for working with employees (Dinero, 2010). It aims to teach the foundations of positive employee interactions, foster teamwork and motivation, and manage conflict effectively in the workplace (Kováčová, 2012). Such a system is necessary to promote a safe and healthy workplace. The themes are discussed throughout the JRT classes utilising actual case studies with a fictitious boss and employee (Huntzinger, 2016). Real-world case studies assist employees in comprehending the situation at hand and streamlining their work processes.

According to Graupp and Wrona (2010), JRs help managers improve their leadership skills by instructing staff on how to avoid personnel difficulties and establishing a standard for excellent relations with all employees. JRs support managing personnel issues effectively and appreciating each employee as an individual. Since this is the case, refer to 'respect for people combined with job instruction', which is at the core of lean construction and is regarded as one of its pillars (Korb, 2016).

Koskela (1992) explained that lean aims to maximise value while minimising waste. The construction industry has shown a growing interest in adopting lean construction. Despite the higher complexity and higher level of uncertainty involved in construction projects compared with the manufacturing industry, lean construction shares common elements with lean manufacturing (Erthal & Marques, 2022). Demirkesen et al. (2022) noted that the benefits of implementing lean principles in the construction industry have previously been indicated as cost savings, time reduction, higher productivity, higher quality, better relations with customers, reduced rework, less waste, less inventory, increased worksite safety, fewer variations in the project, and increased worker motivation.

1.3   Research methods

The case study method enabled the researchers to investigate the TWI phenomenon and identify the factors preventing workers from using it to improve OHS on construction sites. As shown in Table 1.1, multiple case study designs were used in this study's research design. The nature of the selected construction projects was building works. This is because building works expose workers to numerous hazards that must be addressed to reduce accidents on construction sites. The selection of the construction projects demonstrated literal replication as the case studies' results were similar. This highlight that the construction sites in the city of Bloemfontein in South Africa are experiencing challenges concerning poor OHS.

The qualitative research approach was adopted as recommended by Tracy (2013). This is because it assisted the researchers in investigating the OHS issues from the lived experience of the interviewees in the multiple case study projects. The procedure for collecting data was guided by an open-ended question through semi-structured and focus group interviews. Across the multiple case study projects, the researcher collected the data by highlighting the background of the study to the interviewees. This process helped the interviewees to understand the subject under investigation. Thus, the selection of the interviewees adopted purposive sampling because the researchers only interviewed people working on the selected construction projects. This decision is also supported by Tracy (2013), namely that the interviewees must have the necessary knowledge and experience relating to

Table 1.1 Research sample

Case studies	Interviewees	Total (units)	Rate (%)
Building project 1 (BP1)	**Semi-structured interviews**		
	Construction manager (CM)	5	17%
	Safety manager (SM)		
	Site engineer (SE)		
	Junior foreman (JF)		
	Senior foreman (SF)		
	Focus group interviews		
	Group of artisans (GA)	4	43%
	Group of workers (GW)	9	
Building project 2 (BP2)	**Semi-structured interviews**		
	Construction manager (CM)	5	17%
	Safety manager (SM)		
	Site engineer (SE)		
	Senior foreman (SF)		
	Junior foreman (JF)		
	Focus group interviews		
	Group of artisans (GA)	7	23%
Total interviewees		**30**	**100%**

the problem being investigated. In addition, the interviews took place face-to-face between the researchers and the interviewees on the selected construction projects.

Before the focus group interviews, the study's purpose was communicated to each group of interviewees. The interviewees were made aware of the interview's procedure and encouraged to participate in an open dialogue where everyone could express their thoughts without limitation (Yin, 2014). In the case studies, the interviews were recorded and later transcribed to aid analysis. As a result, thematic analysis was adopted to analyse the data collected in the multiple case study projects (Castleberry & Nolen, 2018). In summary, using a thematic analysis provided a platform that helped the researchers to identify a theme (improving work-related safety using TWI). The identified theme enabled the researcher to interpret and explain the data. Data that failed to address a theme or the reported OHS issue were excluded.

1.4   Results

The data analysed from the two case studies indicated that the issues concerning OHS in construction are rooted in the methods implemented by construction team leaders. Construction team leaders oversee a team of workers on a project and typically hold managerial roles. Some interviewees from multiple case studies also support this statement. They explained that the construction project leaders supervise the artisans, general workers, and plant operators to improve construction site productivity and safety management systems. In addition, Cooper (2002) also highlighted that project leaders are responsible for identifying hazards and preventing accidents on construction sites.

1.4.1 Safety training and skills transfer

Regarding safety training (skills transfer) on construction sites, an interviewed building project 1-site engineer (BP1-SE) explained that there are numerous barriers preventing the transfer of skills on construction sites. Most foremen and artisans are under pressure to complete their work as quickly as possible owing to their high levels of performance. Therefore, they regularly disregard the safety regulations and often issue instructions not to follow relevant health and safety standards for the workers to complete their tasks on time. This could be one of the causes of the high accident rate in the construction industry. There is a big gap in promoting safety training on construction sites because senior personnel value productivity over safety. Another difficulty with skills transfer is that some foremen and artisans are apprehensive about sharing their expertise with the workers for fear that workers may someday take over their positions.

It was further reported by some interviewees from BP1 and BP2 that experienced foremen frequently feel threatened when their employers hire graduates from colleges or universities. Given that most of the apprentices want to deploy their new skills to assist them in performing their daily duties, this may be why foremen feel threatened. For example, building project 2-site engineer (BP2-SE) and BP1-JF (junior foreman) indicated that senior foremen (SE) frequently withheld their skills from them in executing their tasks out of fear that they could lose their positions. Such acts have often resulted in tension between workers and foremen. Therefore, it can be concluded that most foremen, also known as mentors to the apprentices, need to acknowledge training by intention. For instance, the apprentices would better understand the significance of adhering to safety standards if they received the appropriate training from the senior workers.

In addition, BP1-CM (contract manager) supported the statement made by BP2-SE that foremen often needed to share their experience with the workers. This is because most foremen believe workers are there to take their jobs. The interviewee went on to say that the foremen needing to compete with the workers in utilising technology tools was the biggest difficulty they were facing. Some foremen and site supervisors are alleged to intimidate and bully the apprentices on the work site. However, an interviewed BP2-SF is quoted as follows, when responding to the claims that they are abusing apprentices:

I often struggle to understand how some university graduates behave; they refuse to put in the extra effort or let us teach them the job. They don't want to try to learn how to do it themselves because they are constantly working on their laptops. They approach us with the mentality that they are educated and there is nothing they will learn from us, uneducated foremen.

> *Additionally, even when projects are computer-designed, the actual work is still done by hand, so some learners find it difficult to understand how labour-intensive the construction industry is. They should understand that they are responsible for assembling the project according to specifications and most of the job is done physically, not behind the computer.*

The lack of job stability among foremen in the construction industry may have influenced the decision-making regarding skills transfer or knowledge sharing among construction workers. As a result, this has influenced the unsafe behaviour of people on construction sites. Failure by foremen to promote safety while completing the task could be the root cause of unsafe behaviour on construction sites. This explains why Awwad et al. (2016) highlighted that occupational accidents, injuries, and illness remain a serious problem in the construction industry.

In addition, a BP2-CM outlined that on his project, he is currently investigating allegations laid against some of his foremen. They have been accused by the apprentices that they need to be given full responsibility and are often side-lined from making decisions and working according to specifications. Such behaviour by the foremen often places apprentices in an awkward position since they were deemed to make mistakes or errors, which might result in loss of production, material waste, and even accidents. Also, an interviewed BP1-SF and BP2-JF explained that foremen are the best people to train new workers on site. This is because most of the critical tasks are completed by the foremen, and this would provide the best platform for apprentices to understand how construction projects are being carried out.

1.4.2   Implementation of safety management systems

Regarding implementing the SMS on construction sites, the BP1-SM and BP2-SM had the same responsibility, namely to design a safety system that would promote a safe and healthy working environment. A BP1-SM further reported that they are also responsible for supervising people in construction to comply with the adopted safety regulations. This is because failure to comply with safety regulations has resulted in many accidents experienced on construction sites.

In addition, most of the interviewees in BP1 and BP2 stated that many companies struggle to provide adequate personal protective equipment (PPE) and to organise training facilities for people on construction sites. A BP2-SM highlighted that there is a facility to train the workers on their project. She often trains the workers under the tree outside her office. Her team was responsible for training all site visitors and workers regarding the safety guidelines established on construction sites. The interviewee continued by saying that they were responsible for ensuring that each worker wore all the necessary PPE when reporting to work. Workers should wear the correct PPE, which includes safety goggles, helmets, gloves, earplugs, safety boots, and suits (overalls), all designed to prevent various incidents.

1.4.3   Indifference to safety rules

In addition, a BP2-CM voiced the same sentiments regarding some of the artisans that are out of favour with the construction management team, especially the construction manager and some foremen. This is because the construction manager and some foremen have developed the tendency to ignore the safety rules to speed up productivity. They argue that high levels of productivity are what brings food to the table, and everyone should work hard even if they must ignore certain rules. This statement highlights the fact that there is a lack of a collaboration between the safety management team and the construction management team. This is because it is reported that site agents often prioritised productivity over the lives of the workers. For instance, they allocated an unreasonable time for the safety team to provide safety training to the workers. Also, they experienced challenges where site agents interfered with their training program by rushing the safety team to finish as quickly as possible. Site agents would argue that the safety team was not finishing their toolbox in time, and this affected them negatively because they were running behind schedule.

The safety management team may choose to prohibit the workers from continuing operations if they have ignored the safety protocols, according to the BP1-SE. However, a site agent would give the workers orders to disregard the recommendations made by the safety team. The site agent routinely intimidated the workers, threatened to fire them, induced fear in them, and pressurised them to disregard safety instructions and laws. One of the people interviewed asserted the following:

> *The site agents must be feared by the workers. The workers are frequently given explicit instructions by the site agents that they are neither being paid nor employed by the safety manager. They must be aware of the real boss and the person who has the authority to fire them.*

Some interviewees reported in BP1 and BP2 that it is the responsibility of the construction management team to manage the project to maintain high levels of productivity. On construction sites, high levels of productivity are promoted by designing an effective construction method that would eliminate waste and continue improving production without placing the lives of people in construction in danger. In addition, construction managers are also responsible for teaching the workers to work according to the designed SMSs. However, this statement of a SE and a CM contradicts the statement issued by a SM in the previous paragraph.

One of the challenges highlighted by a SM concerning the lack of budget allocation for providing adequate training is corroborated by Allen (1919). This issue has existed for decades, and Allen noted that providing training incurred an overhead charge for the organisation. In addition, a BP1-CM explained that contractors need to be allocated more budget to provide safety training for the people on construction. Instead, they only allocated a budget to design the SMS, hire a safety management team and buy PPE clothes.

1.4.4   Learning from on-the-job experience

According to Reese (2011), the purpose of implementing learning by doing is to provide a pathway for new employees to learn from experienced employees compared with learning from watching others perform, reading others' instructions, or listening to others' lectures. This might be referenced to the construction management team, including foremen and artisans because they have gained previous experience. In addition, a foreman explained that they were responsible for teaching workers how to set up work and assemble the project according to specifications. The interviewee further explained that they could complete their task on schedule through teamwork. The new employees had to understand that construction work is dangerous and complicated. Their decision-making abilities may either save a life or create a disaster, resulting in injury, causing accidents, or even causing the death of other workers.

The case study's results showed that workers are vulnerable to human error and violations without adequate training. As has been reported, the unsafe behaviour of people on construction sites could be the source of accidents experienced on sites. The data records challenges experienced in construction by trainees. For example, a BP1-CM said:

> *The construction management team is experiencing challenges relating to teaching people to learn the construction work while most of them are not interested in pursuing careers in the construction industry. He explained that most of the workers are working on his project because there is no other jobs they could work.*
>
> *Another challenge is that most workers report working under the influence of alcohol. This is because they have no respect [for] the nature of construction work, which exposes them to hazards that might result in accidents. Lastly, there are challenges relating to their social life, which might be why most of them find relief and healing in drinking excessively.*

It was found that the construction team members needed help to promote learning-by-doing on project sites. Several challenges were influencing poor training and skills transfer on project sites. The cited challenges include the following:

- Insufficient training budget;
- Knowledge and skills transfer failure;
- Lack of interest in work in the construction industry;
- Social issues affecting workers;
- Construction productivity; and
- Lack of training facilities on construction sites.

The BP1-GW, interviewed through a focus group, explained their challenges and benefits concerning the designed safety system. One of the benefits they highted is that their company was trying hard to comply with the safety standards. Enough

PPE had been provided for everyone working on sites. Also, the safety managers prioritised a safety toolbox talk every week. During these sessions, the safety manager presented the safety risk assessment of each activity they would be working on during the week. Additionally, they disclosed that throughout the week, safety inspectors and representatives would examine them to ensure that they were following the safety risk assessment plan that had been issued for a particular activity.

The interviewees indicated that they frequently encountered difficult situations, particularly when there was tension between the safety and construction management teams. Such altercations indicated that everyone was looking after themselves. Also, such tension between the two sides had a negative effect on them because both were their bosses, and they could not afford to disobey either of them. Tension frequently arose when the foreman requested that they ignore safety rules to finish the assignment by the deadline. The safety manager's claim that personnel were routinely made to disregard safety regulations and rules was supported by this claim. These are some of the challenges to which workers are exposed, with little recourse.

In addition, one of the workers in BP1-GW explained that their employers often do not live up to their promises. He explained that at the start of the project, their construction manager promised them that they would send the best performing workers to training relating to scaffold erection. However, not a single worker had been selected for training. The response they received from the construction manager was that they had to cancel the scheduled training because the company had made a loss and their budget was depleted. They tried to negotiate with the clients who informed them that no training budget had been allocated for the project.

1.5   Discussion

The construction industry has adopted OHS to minimise accidents and create safe and healthy working environments for workers. However, despite the adoption of OHS, the industry continues to record a high number of accidents. For example, the Department of Labour (2017) reported that in 2017, there were 1.5 to 2.5 fatalities per week on average in the South African construction industry. To address such challenges, this study investigated the factors preventing workers from adopting TWI and improving OHS through learning by doing on construction sites. As a result, a case study research approach was adopted to help identify factors preventing workers from using TWI to enhance OHS through learning by doing on construction sites.

The case study's results showed that more funding is needed for the safety management teams to provide workers with adequate training on construction sites. Additionally, it should be mentioned that contractors had used their income to cover overhead expenses for OHS training. This is because contractors did not receive funding for workers OHS training. Instead, they only set aside funds to hire a safety management team, design a SMS, and purchase PPE clothing.

Additionally, it was found that the interviewed construction team members needed help to promote learning by doing on construction sites owing to various

Insufficient training
budget

Lack of training
facilities

Knowledge and skill
transfer failure

**Inadequate
training and skill
transfer**

Construction
productivity

Lack of interest in
working in the
construction industry

Social issues affecting
workers

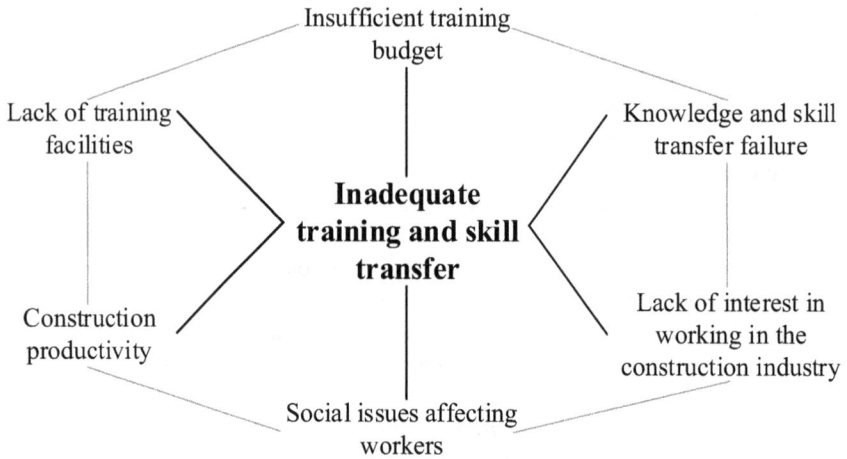

Figure 1.1 Factors contributing negatively to the implementation of learning by doing on
construction sites

OHS challenges. On construction sites, several issues led to substandard training
and skills transfer. Figure 1.1 summarises the factors that contributed to the need
for more implementation of learning by doing to improve OHS on construction
sites. It should be highlighted that a lack of implementation of learning by doing
is mostly caused by inadequate training and skills transfer. Inadequate training
and skills transfer are also influenced by other factors, such as insufficient training budget and knowledge, skills transfer failure, lack of interest in working in the
construction industry, social issues affecting workers, construction productivity,
and lack of training facilities.

The philosophy of the TWI method is outlined by Huntzinger (2016). The
author indicated that the TWI method's philosophy promotes learning by doing,
which means handling problems under the guidance of an appropriately competent supervisor. Therefore, supervisors must understand how workplace training
impacts workers. Most of the knowledge and skills workers utilise to perform their
duties across the construction industry were acquired on the job through learning by doing. Therefore, contractors must mitigate the identified factors contributing negatively to the implementation of learning by doing on construction sites
(Figure 1.1). In addition, the client must support the contractors to encourage learning by doing, which would help to improve the industry's subpar OHS.

The case study's results also showed that SMs and CMs regularly disagree
with one another. Conflicts arise because the SMs frequently endanger the workers' lives to increase productivity. Additionally, under the strain of their duties, the
site managers may compel workers to neglect safety protocols. According to Allen
(1919), the notion that a man qualified for the position is also qualified to serve as a

supervisor at work is sometimes somewhat misguided, as highlighted in the previous statement. It can be concluded that not all managers and supervisors possess the knowledge necessary to guide or instruct their teams at work. Indeed, management frequently endangers workers' lives in the construction industry through their actions or decisions.

1.6   Chapter summary

The case study's results support the notion that the construction industry must continue improving OHS. Thus, this chapter identified the factors preventing workers from using TWI to OHS through learning by doing on construction sites. However, it was found that the members of the construction teams need help to encourage learning by doing, which is the philosophy of the TWI method. Lack of implementation of learning by doing is mostly caused by inadequate training and skills transfer, as highlighted in Figure 1.1. Management will gain from this by encouraging learning by doing on their construction sites. The learning-by-doing effect on construction sites will help workers improve their work practices regarding health and safety in the construction industry. It will also help new workers or apprentices to learn from experienced workers. Future studies need to be conducted to understand the function that TWI would play in establishing guidelines for improving construction practices and encouraging learning on construction sites.

References

Ajslev JZ and Nimb IE (2022) Virtual design and construction for occupational safety and health purposes – A review on current gaps and directions for research and practice. *Safety Science* 155: 105876. Available from: https://doi.org/10.1016/j.ssci.2022.105876

Allen E (1919) *The instructor. The man, and the job.* Philadelphia and London: J.B. Lippincott Company.

Awwad R, Souki OE and Jabbour M (2016) Construction safety practices and challenges in a Middle Eastern developing country. *Safety Science* 83: 1–11.

Castleberry A and Nolen A (2018) Thematic analysis of qualitative research data: Is it as easy as it sounds? *Currents in Pharmacy Teaching and Learning* 10: 807–815.

Cooper D (2002) Human factors in accidents. In: *Revitalising health and safety-achieving the hard target.* (pp. 1–7). Available from: https://www.behavioral-safety.com/articles/Human_factors_in_accidents.pdf (accessed January 02, 2023).

Demirkesen S, Sadikoglu E and Jayaman E (2022) Investigating effectiveness of time studies in lean construction projects: Case of Transbay Block 8. *Production Planning & Control* 33(13): 1283–1303. Available from: https://doi.org/10.1080/09537287.2020.1859151

Department of Labour (2017, March 9) *Labour on injuries and fatalities in SA construction sector.* Pretoria, Gauteng Province, South Africa. Available from: www.gov.za/speeches/sa-construction-sector-9-mar-2017-0000 (accessed February 14, 2020).

Dinero D (2010) *Training within industry: Fundamental skills in today's workplace.* New York: The TWI Learning Partnership Rochester.

Erthal A and Marques L (2022) Organisational culture in lean construction: Managing paradoxes and dilemmas. *Production Planning & Control* 33(11): 1078–1096. Available from: https://doi.org/10.1080/09537287.2020.1843728

Feng PP and Ballard G (2012) Standard work from a lean theory perspective. In: *Proceedings for the 16th annual conference of the international group for lean construction* (pp. 703–712). Manchester: IGLC.

Georgise FB and Mindaye AT (2020) Kaizen implementation in industries of Southern Ethiopia: Challenges and feasibility. *Cogent Engineering* 7(1): 1823157. Available from: https://doi.org/10.1080/23311916.2020.1823157

Graupp P and Wrona RJ (2010) *Implementing training within industry – Creating and managing a skills-based culture*. London and New York: CRC Press.

Grip AD and Sauermann J (2013) The effect of training on productivity: The transfer of on-the-job training from the perspective of economics. *Educational Research Review* 8: 28–36.

Health and Safety Executive (HSE) (2019) *Kinds of accident statistics in Great Britain, 2019*. London: The Health and Safety Executive. Available from: www.hse.gov.uk/statistics/causinj/kinds-of-accident.pdf

Huntzinger J (2006) Why standard work is not standard: Training within Industry provides an answer. *Association for Manufacturing Excellence* 22(4): 7–13.

Huntzinger J (2016) *The roots of lean: Training within industry: The origin of Japanese management and kaizen and other insights*. Indianapolis, IN: Lean Frontiers, Inc.

Korb S (2016) Respect for people and lean construction: Has the boat been? In: Pasquire CA (Ed.), *Proceedings 24th annual conference of the international group for lean construction* (pp. 43–52). Boston, MA: International Group for Lean Construction.

Koskela L (1992) *Application of the new production philosophy to construction*. Finland: Centre for Integrated Facility Engineering (CIFE).

Kováčová L (2012) The renaissance of method TWI – Training within industry. *Transfer inovácií* 23: 289–291.

Kumar R (2019) Kaizen a tool for continuous quality improvement in Indian manufacturing organization. *International Journal of Mathematical, Engineering and Management Sciences* 4(2): 452–459. Available from: https://doi.org/10.33889/IJMEMS.2019.4.2-037

Latijnhouwers C and Berendsen G (2014) *Training within industry – Job instruction*. Ocheten, Netherland: TWI Institute Netherland.

Liu R, Liu Z, Liu H-C and Shi H (2021) An improved alternative queuing method for occupational health and safety risk assessment and its application to construction excavation. *Automation in Construction* 126: 103672. Available from: https://doi.org/10.1016/j.autcon.2021.103672

Liu R, Mou X and Liu H-C (2020) Occupational health and safety risk assessment based on combination weighting and uncertain linguistic information: Method development and application to a construction project. *IISE Transactions on Occupational Ergonomics and Human Factors* 8(4): 175–186. Available from: https://doi.org/10.1080/24725838.2021.1875519

Mahmoudi S, Ghasemi F, Mohammadfam I and Soleimani E (2014) Framework for continuous assessment and improvement of occupational health and safety issues in construction companies. *Safety and Health at Work* 5(3): 125–130. Available from: https://doi.org/10.1016/j.shaw.2014.05.005

Miron L, Talebi S, Koskela L and Tezel A (2016) Evaluation of continuous improvement programmes. In: Pasquire C (Ed.), *Proceeding 24th annual conference of the international group for lean construction* (pp. 23–24). Boston, MA: IGLC.

Misiurek K and Misiurek B (2017) Methodology of improving occupational safety in the construction industry on the basis of the TWI program. *Safety Science* 92: 225–231.

Reese H (2011) The learning-by-doing principle. *Behavioral Development Bulletin* 11: 1–19.

Tracy S (2013) *Qualitative research methods: Collecting evidence, crafting analysis, communicating impact*. West Sussex: Wiley-Blackwell.

Yin R (2014) *Case study research: Design and methods*. Los Angeles: Sage Publications.

2 Causes of human failure on construction sites

2.1 Introduction

Occupational health and safety (OHS) is just one of many elements of the workplace that must ensure a physically and psychologically suitable working environment that can produce positive business outcomes (Karanikas et al., 2022). Construction workers in most countries are concerned about the state of OHS (Manu et al., 2017). The OHS of construction workers has long been a significant global issue (Zhang et al., 2020). Many occupational accidents are due to poor OHS implementation (Andersen et al., 2018). For example, it has been reported that more than 60,000 workers worldwide lose their lives because of occupational accidents in the construction industry (Golizadeh et al., 2018). In addition, the Health and Safety Executive HSE; HSE, 2018) reported that between 2017 and 2018, there were 1.64 fatal injuries per 100,000 employees in the construction industry in the United Kingdom (UK). In the United States of America (USA), the situation is no different with the construction industry reporting 19.1% of occupational fatalities in 2016 (Golizadeh et al., 2018).

The above statistics show that occupational accidents are a serious safety concern in the construction industry. It is well known that human failure, in the form of errors, violations, and systematic failure (Reason, 2000), is to blame for most of the construction industry's recorded accidents (HSE, 2009). These causes are due to human factors unique to each worker, such as poor physical and mental health, a lack of knowledge, a negative attitude, and a lack of skills, which are all behavioural factors (Wong et al., 2019). Systemic failure may be tied to the actual working circumstances that the employees face, such as a hazardous work environment and outdated equipment (Wong et al., 2019). Therefore, the organisation needs to identify, research, and monitor human factors to control risk because human failure causes accidents (HSE, 2009). This is consistent with the view that learning from past failures is crucial for preventing accidents or injuries (Pan et al., 2022).

Accidents are not limited to developed countries alone. They are also common in South African construction. For example, some of the recorded accidents in the South African construction industry are shown in Table 2.1, where human failure may have played a significant role. Based on this assumption, this chapter highlights the factors causing human failure on construction sites. In this chapter, the

DOI: 10.1201/9781003365341-2

Table 2.1 Examples of construction accidents in South African construction

The details of accidents	The impact of accidents	Source
Durban building collapses on trucks. Date: March 28, 2018. Place: Jacobs in Durban (KwaZulu-Natal Province)	Three employees were killed when a two-storey building, which was under construction, collapsed onto a truck carrying tiles.	Singh (2018)
The temporary M1/Grayston Drive pedestrian and cyclist structural bridge collapsed. Date: October 14, 2015. Place: Sandton in Johannesburg (Gauteng Province)	The collapse of the temporary pedestrian bridge resulted in the death of two employees and 19 others injured as the collapsed structure crushed their vehicles.	Lefifi (2016)
Collapsed building in Meyersdal Eco-Estate. Date: August 18, 2014. Place: Alberton in Johannesburg (Gauteng Province)	Seven employees died and eight others were injured when a large slab of concrete on the upper level of the house reportedly collapsed, causing the ground floor structure to give in under its weight.	Petterson (2014)
The Tongaat Mall collapses Date: November 19, 2013. Place: Tongaat in Durban (KwaZulu-Natal Province)	Two employees died, and 29 others were injured when a mall section collapsed during construction.	Evans (2016)

construction projects were the focus of the human failure investigation because it has been contended that observing workers routinely while they are at work could help uncover the behaviours that contribute to failure in the industry (Noroozi et al., 2013). The mechanism through which unsafe behaviours of construction workers develop during stressful situations must be uncovered to regulate these behaviours effectively and reduce construction accidents (Liang et al., 2022).

2.2   Human failure in industrial workplaces

An accident is an unanticipated incident with an unpleasant outcome and is frequently caused by human activity rather than natural occurrences (Hollnagel, 2004). Awwad et al. (2016) concurred with Hollnagel (2004) that accidents are unpleasant situations that happen unexpectedly to people at work and frequently result in either minor or significant injuries or even fatalities. Additionally, because of the intricate relationships between human (individual) and non-human (organisational) systems, many elements contribute to accidents (Holden, 2011) These are also described as the results of an intentional shift in organisational behaviour brought on by the drive to be cost-effective in a cut-throat, competitive market (Dekker, 2011).

According to Dekker (2014), encouraging people to be watchful or attentive at work does not prevent accidents; instead, it fully displays their ignorance of the real issue. Because there are instances of failures that directly result in incidents and accidents on construction sites, the human factor is pivotal in the construction sector (Yorio & Wachter, 2014). Accidents in the construction industry can be linked to human behaviours related to tasks carried out during construction operations (Misiurek, 2016). The direct cause of many accidents in construction is the result of people's unsafe acts. Moreover, these acts can be connected to the systematic failures that are removed in terms of time or position from the incidents (Manu et al., 2017). Reason (2008) explained the causes of unsafe acts in the industries by stating that an act does not need to be either an error or violation; it can, nevertheless, turn out to be unsafe for the workers on the sites.

As a result, unsafe acts and conditions are closely linked to accidents that stem from errors, violations, and system failure (human failure). The investigation of human failure in all industries has always been part of psychological engineering studies (Isaac, 2002). The background of human failure investigation studies could be traced to the late 1950s and 1960s, when formal methods for identifying and classifying human error in nuclear and military development systems were developed along with hardware reliability approaches (Isaac, 2002). The studies dated back to the late 1950s after the World War period and were prioritised to solve errors in the nuclear and military industry (Baziuk et al., 2016).

Notably, all industries' employees experience human failure regularly (Chiu & Hsieh, 2016). Human failure is described as an individual's improper acts or decisions made at an improper time and location on the job (Chiu & Hsieh, 2016). In addition, people's unplanned actions or judgments might result in workplace events and accidents (HSE, 2009). Human failure was divided into two types by Reason (2008): active failure (errors) and latent failure (violations). People in the workplace tend to make mistakes or take shortcuts that frequently lead to failure, which is one of the factors causing errors and violations (Baziuk et al., 2016).

Reason (2008) contended that workers should be viewed as potentially hazardous or dangerous to the industry. As mentioned above, people at work frequently contribute to human failure. The word 'hazardous' refers to modules with unsafe behaviours related to the most catastrophic workplace errors (Reason, 2008). It is common knowledge that the construction industry does not engender a pleasant workplace, contributing to systemic failure due to the procedures adopted to deliver projects on time and at the lowest possible cost. This is because construction work is conducted in a hazardous environment, which exposes workers to accidents in various circumstances that can cause either minor or serious injuries as well as fatalities (Asanka & Ranasinghe, 2015).

Notably, the main issue facing the health and safety (H&S) of the construction industry is the prevalence of occupational diseases, fatalities, and injuries within the industry (Manu et al., 2017). Individual accidents and organisational accidents are the two main categories of accidents in an industry, according to Reason (2000). Individual accidents can be linked to specific people, whereas organisational accidents can be linked to specific organisations. Because construction projects are designed and carried out by people acting as representatives of their organisations, individual and organisational accidents could be combined and should be investigated jointly (Reason, 2008).

2.3   Research methods

To learn more about the factors that lead to human failure on construction sites, a case study research was adopted as recommended by Tracy (2013). Using this approach enables authors to adopt qualitative data analysis through observation and interaction with the participants about their real-life environment. In this instance, the research design allowed authors to conduct on-site observations (such as visiting construction sites) and interviews with construction team members to understand the causes of human failure better. Therefore, a participant observation and interview protocol helped the researchers in Bloemfontein, South Africa, to investigate the causes of human failure on construction sites.

The data were collected through focus groups, semi-structured interviews, and participant observation. The interviewed participants for this study are detailed in Table 2.2. The authors carefully selected the participants to be interviewed because they wanted to interact with those who had first-hand knowledge of the phenomenon (Maxwell, 2013). Because semi-structured interviews were used, the authors were able to prepare the questions in advance. Additionally, the procedure allowed for open-ended questions, which enhanced the data's depth and reliability for analysis (Maxwell, 2013). Focus group interviews included two groups of general workers and a group of artisans. Two groups of general workers were interviewed; the first comprised four workers, and the second, five workers. For the artisans, there was a single group interview with four workers. As Tracy (2013) recommended, the participants in a focus group interview were given an introduction to the study's topic and urged to give each other space to speak freely without interruption. The interviews took place during lunchtime and lasted between 15 and 30 minutes. Each interview was recorded using a cell phone, and the transcript was then typed up. The qualitative data were thematically analysed, and a total of 18 interviews were conducted.

Table 2.2 Demographic profile of the participants

Interviewees	Position	Number
CSI 1	Construction manager	1
CSI 2	Safety manager	1
CSI 3	Site engineers	1
CSI 4	Foremen	2
CSI 5	Group of nine workers	9
CSI 6	Group of four artisans or skilled workers	4
Total interviewees		**18**

Furthermore, participant observation was used to investigate the behavioural acts of workers on construction sites (Maxwell, 2013). The authors were able to learn about the practices of those engaged in construction site work through participant observation. A site project was routinely observed between July 2018 and November 2018, during which time visits were paid to the site.

2.4   Results

The objective of this study was to answer the research question, 'What are the causes of human failure on construction sites?' According to the results, human failure can be defined as an unsafe man-made event, an incident brought on by a failed plan of action developed by people or an organisation to finish a task or an activity on the site. According to the literature section/review, there are three types of human failure: errors, violations, and systematic failure (Reason, 2000). An error is defined as unintended actions or decisions that are deviations such as stumble, slip or lapse, and mistakes (Reason, 2008). Violations are any traceable deliberate deviations from the rules, procedures, instructions, or regulations (HSE, 2009). Systematic failure is a type of repeatable failure that can occur in even the best organisations. Errors in the organisation's design, operation, production process, installation, or maintenance are typically responsible for this failure (Reason, 2000).

2.4.1   *On-site description of human failure*

Some interviewees also stated that human failure is perceived as an uncontrolled incidence leading to an accident, which could result in worker injuries or fatalities. In addition, interviewee CSI 2 also mentioned how human failure still occurs on construction sites. This is because failure in the construction industry is often the consequence of the behaviour and actions of individuals, both of which are very difficult to anticipate on a construction site. Such failure may be related to errors and violations because safety depends on how workers behave and carry out their duties while working on construction sites. A statement by CSI 2 supports Dekker's (2014) claim that failure stems from unpleasant surprises and may be linked to the erratic actions of some unreliable workers. From the findings, human failure is defined, according to Reason (2008), as either errors or violations created in the operation of site activities. Some interviewees are quoted as follows.

> *The results of human failure in construction are mostly created by the decision-making of the workers. Most of the workers are lazy, ignorant, and negligent and do not follow the safety rules of the sites. For instance, I always complain about workers who like to throw equipment or materials to the ground, especially does who are working at height on scaffolds. I always ask them if they understand what would happen to a scaffold should a fixed coupler be hit by a heavy material or equipment thrown to the ground (CSI 1).*
>
> *It is surprising that all the accidents happening on the site are often blamed on us the workers because we work under the instruction of our*

supervisors. Our supervisors are making our lives hell there are times when we are instructed to put our lives in danger due to productivity. Also, there are times when we are working from Monday to Saturday. You can imagine how our mind and our body are reacting to work pressure and this is one of the reasons why we are making mistakes because of fatigue (CSI 5).

2.4.2   Extent of human failure on sites

Additionally, an interviewee, CSI 1, stated that human failure is a widespread issue that the construction industry is working tirelessly to address. For instance, one of the largest construction companies in South Africa is the subject of a current investigation. The company oversaw the erection of one of the biggest soccer stadiums in Africa and the continent's first super train. The investigation is centred on the collapse of a temporary pedestrian bridge in Johannesburg around the end of 2015. This resulted in two fatalities and 19 injuries. The interviewee emphasised that this bridge collapse demonstrates the major problem with construction safety in the South African construction industry. Nevertheless, despite having all the resources required to create and operate a comprehensive SMS, an organisation may still suffer an accident because it cannot govern how people behave and make decisions at work. The following are quotes from the interviewee:

If such a reputable firm could be a victim of a disastrous construction accident resulting in both fatalities and injuries, we should start asking ourselves questions 'How are small firms with fewer resources when compared to big firms, handling factors causing accidents on their construction sites'.

2.4.3   Human factors influencing accident causations on site

In terms of human behaviour, it is observed that the extent of people's behaviour in their everyday activities can have a direct and immediate effect on the health, safety, and well-being on construction sites. This is due to the assumption that up to 80% of accidents may be caused by workers' actions or inactions (HSE, 2009). Additionally, most respondents, particularly CSI 5 and CSI 6, stated that poor housekeeping is one of the human factors contributing to accidents on construction sites. While they acknowledged that there are occasions when they operate under pressure, they also admitted that they occasionally find it difficult to maintain a clean workplace and store waste in the proper location. This reported failure is related to systemic failure because organisations continue to maintain inadequate housekeeping even if they are aware that doing so undermines the goal of ensuring a safe and healthy workplace. Additionally, higher production is associated with yet another systematic failure. Organisations usually find themselves unable to assign one of their members to clean the site because of the productivity targets they must reach. The client's top priority is productivity, and any deviation will result in financial penalties. The interviewees said that anyone visiting the site can see proof that the people working there do not prioritise housekeeping.

One of the interviewees is also quoted as stating the following:

There are times when our project supervisors are instructing us to proceed with the next work activities without cleaning the site activities we worked on. The supervisors would instruct us that our goal is to finish the next activity as soon as possible and we will worry about cleaning the site later.

The majority of those interviewed added that human failure is caused by workers' negligence, ignorance, and disregard for safety regulations in the construction industry. According to CSI 4, human failure is caused by decisions made throughout operations. The interviewee emphasised that the inherent difficulty and unpredictability of construction work are two of the causes of accidents. On construction sites, decisions are made that have an impact on and frequently save the lives of individuals who work there. Other times, though, their decisions frequently result in unsafe behaviour that could endanger lives through accidents. It has been argued that the way construction activities are conducted makes it easy for workers to make mistakes. Further information was provided by CSI 3, who stated that those who take a shortcut or are too lazy to follow the proper method frequently make mistakes. The interviewee said that most people in the construction industry develop a culture in which bad choices, such as taking a shortcut, eventually become acceptable. Most errors can be attributed to management and workers, as they frequently disregard safety protocols, thereby fostering a culture where wrong is considered right until an incident or accident occurs. Errors are said to either be knowledge based or rule based (Reason, 2008). Rule-based errors are linked to organisational or management decisions, while errors that are knowledge-based are linked to individuals' decisions.

2.4.4 Instructions to expedite unsafe actions on sites

According to CSI 5, they frequently receive instructions to carry out tasks in a way that endangers their lives. There are times when construction managers will order them to act without first assessing risk. For instance, they frequently receive instructions to use an unsafe scaffold. These are deliberate actions that violate established OHS regulations. In addition, the workers also complained that they needed to receive appropriate safety training and were frequently given instructions without being allowed to voice their ideas. Another issue mentioned is the lack of sufficient personal protective equipment (PPE) that would last for the project's duration. The safety representatives always complain that they need more budget allocated for safety clothing. This acts as a system failure because it is a recurrent error trap on construction sites and the construction processes that give rise to them.

Furthermore, CSI 1 mentioned that the representatives of the consulting engineers make his job challenging. When visiting the engineer's site, the representatives refuse to wear complete PPE. He indicated that:

There is a time when I was told that they are going charge or give a penalty to my firm or might delay processing a payment certificate. The reason why I received this threat is that I threatened to kick out a chief engineer from visiting my site. The chief engineer was refusing to wear a safety helmet, he complained that the sun is too hot and there is no way he will put on the helmet on his head because he is going to have a headache and will not be able to work for the rest of the day.

When compared with the literature, the chief engineer's behaviour described above can be regarded as undesirable and improper in the construction industry. Images captured on construction sites that identify a few violations that can cause accidents are shown in Figure 2.1. Although it is obvious that no one intended to be

Figure 2.1 On site observation of human factors

harmed or put at risk by these violations, they could result in blunders that cause unplanned accidents.

For example, the first picture from the top left in Figure 2.1 shows the excavated trench, which had yet to be barricaded. Individuals visiting or working on a site might fall into a trench – an error resulting in a minor or major injury. As illustrated in the literature, such violations could be linked to a failed or poor working procedure (HSE, 2009). The second picture from the top left shows an incompletely erected scaffold. Furthermore, no warning sign indicates whether the scaffold is ready to be used. This violation puts the workers in danger because if a worker climbs on such a scaffold and is injured, the organisation might be charged by the Department of Labour for failure to warn the workers about the condition of a scaffold.

The top right and centre left pictures in Figure 2.1 show the outcomes of poor housekeeping, which relate to system failure. Poor housekeeping can be seen as loose materials such as mortar, short steel reinforcement, bricks, roof purlins, and scaffold tubes scattered throughout the workplace. The removal and safe storage of these materials and equipment have become routine tasks for construction workers and project supervisors, and according to Reason (2008), a routine task is more prone to omission because our minds are on the next activities. Additionally, it has been noted that poor housekeeping is a factor that could lead to minor or major accidents on the site.

The image in the centre depicts a water leak from a pipe that supplies water to various workstations. Unsafe working circumstances are caused by pooled water that may result in errors such as slips, trips, and falls. Water leaks are an unintentional violation. This breach occurs due to insufficient resources, including a broken water pipe, and raises concerns about the standard of supervision at work.

2.5   Discussion

Regarding PPE, it has also been noted that most workers, even consultants' representatives, frequently need to wear full PPE. During an interview, a construction manager questioned this claim. Figure 2.1, however, does not contain images of people who are not fully protected. Furthermore, there was no signage alerting people to the excavation of trenches, the use of scaffolds, or the installation of steel reinforcing on the property. According to the site layout plan, there is neither a pedestrian nor a vehicle road. People who are working or visiting a location share the road with vehicles. These violations result from inadequate project supervision, a lack of accountability, and undisciplined behaviour on the part of construction site workers. According to Fernández-Muñiz et al. (2017), the environment in which individuals work impacts their behaviour. For instance, Figure 2.1 depicts several errors, violations, and system failures during the site project observation.

The interviewees draw attention to the fact that blaming someone for human failure in the construction industry is now regular practice. It should be noted that the interviewees attributed blame for the human failure to workers, management, and construction projects. For example, focus group interviewees linked human

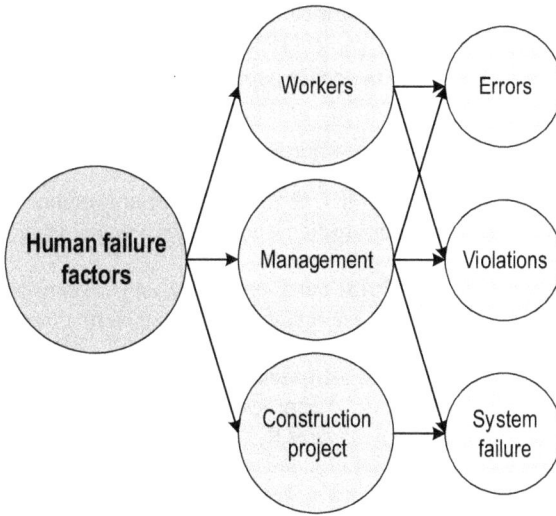

Figure 2.2 Factors causing human failure on construction sites

failure to management's decisions and actions, whereas the construction managers blamed it on workers' failure to comply with the safety regulations. Some of the workers and artisans interviewed blamed the human failure on the nature of the construction project. The identified causes of failures on construction sites are depicted in Figure 2.2. It should be observed that human failure is connected to the worker, management, and construction project factors. The workers' factor relates to human failure because of errors and violations. The management factor is related to human failure because of errors, violations, and system failure, while the construction project factor is related to human failure because of system failure.

2.6   Chapter summary

OHS is a problem in the construction industry. Human failure, whether it takes the form of errors, violations, or system failure, is undesirable because it may contribute to accidents. Table 2.1 lists a few instances of accidents that significantly contributed to South African construction accident statistics. The findings indicate that human failure is defined as an unsafe human-made occurrence brought on by unsuccessfully planned or executed activities, either by people or by an organisation. The causes of human failure are blamed on workers, management, and construction projects. In essence, failure is largely due to human behaviour that is associated with workers and management. A construction manager who complained about a chief engineer who would not put on full PPE when visiting a site supports this assertion. No one is permitted access to a work site unless they are fully protected from the environment, as stated in the construction regulations in South Africa. Similar regulations can be found in several developed and developing countries.

The behaviour of the cited chief engineer cannot be ignored as it sets a wrong and dangerous precedent. As a result, project stakeholders should increase enforcement and inspectorate capabilities in the construction industry to make worksites safer.

References

Andersen LP, Nordam L and Joensson T, et al. (2018) Social identity, safety climate and self-reported accidents among construction workers. *Construction Management and Economics* 1: 22–31.

Asanka WA and Ranasinghe M (2015) Study on the impact of accidents on construction projects. In *Proceedings of the 6th international conference on structural engineering and construction management* (pp. 58–67). Kandy, Sri Lanka: CSECM.

Awwad R, Souki OE and Jabbour M (2016) Construction safety practices and challenges in a middle eastern developing country. *Safety Science* 83: 1–11.

Baziuk P, Núñez Mc Leod J, Calvo Olivares R, et al. (2016) Modelling human reliability: The underlying cognitive abilities. In *Lecture notes in engineering and computer science*. London: International Association of Engineering.

Chiu MC and Hsieh MC (2016) Latent human error analysis and efficient improvement strategies by fuzzy TOPSIS in aviation maintenance tasks. *Applied Ergomics* 54: 136–147.

Dekker S (2011) *Drift into failure: From hunting broken components to understanding complex systems*. Surrey: Ashgate Publishing.

Dekker S (2014) *The field guide to understanding "human error"*, 3rd ed. London & New York: CRC Press, Taylor & Francis Group.

Evans J (2016, May 5) *Tongaat Mall collapse report to be given to NPA*. Available from: www.news24.com/SouthAfrica/News/tongaat-mall-collapse-report-to-be-given-to-npa-20160524

Fernández-Muñiz B, Montes-Peón JM and Vázquez-Ordás CJ (2017) The role of safety leadership and working conditions in safety performance in process industries. *Journal of Loss Prevention in the Process Industries* 50: 403–415.

Golizadeh H, Hon CKH, Drogemuller R and Hosseini MR (2018) Digital engineering potential in addressing causes of construction accidents. *Automation in Construction* 95: 284–295.

Health and Safety Executive (HSE) (2009) *Reducing error and influencing behaviour*. London: Health and Safety.

Health and Safety Executive (HSE) (2018) *Occupational lung disease in Great Britain*. London: Health and Safety Executive.

Holden PJ (2011) People or systems? To blame is human. The fix is to engineer. *Primary Care (PMC)* 54(12): 34–41.

Hollnagel E (2004) *Barriers and accident prevention*. Aldershot: Ashgate Publisher.

Isaac A (2002) Human error in air traffic management: Deviation or deviance. In: McCabe P (Ed.), *Contemporary ergonomics* (pp. 230–236).

Karanikas N, Mohammad S and Hasan T (2022) Occupational health & safety and other worker wellbeing areas: Results from labour inspections in the Bangladesh textile industry. *Safety Science* 146:105533. Available from: https://doi.org/10.1016/j.ssci.2021.105533

Lefifi T (2016) *Who's to blame for Grayston Bridge collapse?* Johannesburg, RSA: City Press. Available from: http://city-press.news24.com/Business/whos-to-blame-for-grayston-bridge-collapse-20160710

Liang Q, Zhou Z, Ye G, et al. (2022) Unveiling the mechanism of construction workers' unsafe behaviors from an occupational stress perspective: A qualitative and quantitative examination of a stress-cognition-safety model. *Safety Science* 145: 105486. Available from: https://doi.org/10.1016/j.ssci.2021.105486

Manu P, Gibb A, Manu E, et al. (2017, April) The role of human values in behavioural safety. Institution of civil engineers. In McAleenan A (Ed.). *Proceedings of the Institution of Civil Engineers* 170(MP2): 49–51.

Maxwell J (2013) *Qualitative research design: An interactive approach*, vol. 3. Los Angeles, CA: Sage Publishing.

Misiurek B (2016) *Standardized work with TWI: Eliminating human errors in production and service processes*. New York: CRC Press.

Noroozi A, Khakzad N, Khan F, et al. (2013) The role of human error in risk analysis: Application to pre and post-maintenance procedures of process facilities. *Reliability Engineering and System Safety* 119: 251–258.

Pan X, Zhong B, Wang Y, et al. (2022) Identification of accident-injury type and bodypart factors from construction accident reports: A graph-based deep learning framework. *Advanced Engineering Informatics* 54: 101752. Available from: https://doi.org/10.1016/j.aei.2022.101752

Petterson D (2014, August 20) *Collapsed building illegal*. Available from: www.infrastructurene.ws/2014/08/20/collapsed-building-illegal/

Reason J (2000) Human error: Models and management. *British Medical Journal* 320: 768–770.

Reason J (2008) *The human contribution: Unsafe acts, accidents and heroic recoveries*. New York: Routledge.

Singh K (2018, March 28) *Three die after Durban building collapses on truck*. Available from News24: https://www.news24.com/SouthAfrica/News/developing-two-trapped-after-durban-building-collapses-on-truck-20180328

Tracy S (2013) *Qualitative research methods: Collecting evidence, crafting analysis, communicating impact*. Chichester, West Sussex: Wiley-Blackwell.

Wong FK, Chiang Y, Abidoye FA, et al. (2019) Interrelation between human factor – Related accidents and work patterns in construction industry. *Journal of Construction Engineering and Management* 145(5).

Yorio P and Wachter JK (2014) The impact of human performance focused safety and health management practices on injury and illness rates: Do size and industry matter? *Safety Science* 62: 157–167.

Zhang M, Shi R and Yang Z (2020) A critical review of vision-based occupational health and safety monitoring of construction site workers. *Safety Science* 126: 104658. Available from: https://doi.org/10.1016/j.ssci.2020.104658

3 Human failure and safety management systems

3.1 Introduction

Improper work procedures negatively impact the health, safety, and well-being (HSW) of workers in construction. This is because of the construction work's complexity and intractable nature, which emphasises adopting an effective safety management system (SMS; Ranasinghe et al., 2020). The lack of effective SMSs in construction has contributed to many accidents (Chi & Lin, 2022). In most countries, construction operations are responsible for unfortunate incidents of industrial accidents (Koc & Gurgun, 2022). For example, from 2015 to 2021, there were 54 964 injuries on construction sites in South Africa that needed medical attention (Federated Employers Mutual Assurance Company [FEM], 2022).

Accidents are a nightmare for people in construction despite implementing the SMS. As a result, the SMS has remained committed to the Safety-I approach that focuses on learning from negative accidents or events. Hollnagel (2018) defined Safety-I as a state where as little as possible goes wrong in a system. Safety-II is described as an approach that emphasises ensuring that as many things as possible go smoothly in the system (Hollnagel, 2018). Nevertheless, the inadequate use of Safety-II, which emphasises enhancing safety through an analysis of daily work, remains a challenge in the construction industry (Martins et al., 2022). Moreover, Safety II does not replace Safety-I (Woodward, 2019). According to Provan et al. (2020), SMS in Safety-I aims to align and control the organisation and people in construction (PiC) by indicating what is safe, whereas SMS in Safety-II aims to enable the organisation and its people to adapt safely to emergent situations and conditions. Both Safety-I and Safety-II are in accordance with SMS because they aim to reduce occupational risk. Nevertheless, they could also expose an organisation to more hazards (Hollnagel, 2018).

Poor SMS and human failure in construction work are so closely linked that one can conclude that the latter is not a coincidence (Health and Safety Executive [HSE], 2009). Human failure is categorised into errors and violations (Reason, 2000), and they are all associated with Safety-I and Safety-II. For example, a slip or lapse is an error that defines unintended or unplanned conduct (Reason,

DOI: 10.1201/9781003365341-3

2016). Errors could be called Safety-I because it focuses on learning from negative accidents related to slips or lapses. The term 'violation' refers to intentional behaviour such as non-compliances, circumventions, shortcuts, and workarounds (HSE, 2019). Safety-I may be relevant because it focuses on understanding accidents caused by non-compliance, circumventions, shortcuts, and workarounds. Additionally, errors and violations are relevant to Safety-II because errors and violations continue to occur on construction sites despite the implementation of SMS. Safety-II is based on continuous learning, which enables the recording and comparing daily discussions about the work-as-done with the work-as-imagined (Martins et al., 2022).

Previous studies have indicated that accidents involving human factors occur in various industries (Ghasemi et al., 2022). Eliminating every aspect that might affect how individuals behave negatively at work is a step towards better HSW. In the mode of Safety-II, PiC can change the workplace environment to make it safer (Xia et al., 2021). However, how poor SMS could contribute to accidents through human failure is illustrated in Figure 3.1, which presents the classification of causes of accidents. Poor SMS could result in human failure in the form of errors and violations. Errors and violations in industrial workplaces relate to Safety I and Safety II through workers' behaviour and organisational systems. The impact of errors and violations often results in injuries and fatalities.

To solve this reported problem of human failure that results in accidents as highlighted in Figure 3.1, this chapter presents mitigation measures. According to Chi and Lin (2022), SMS (including organisational structures, accountability, roles, rules, and procedures) is a systematic approach to managing safety. Construction

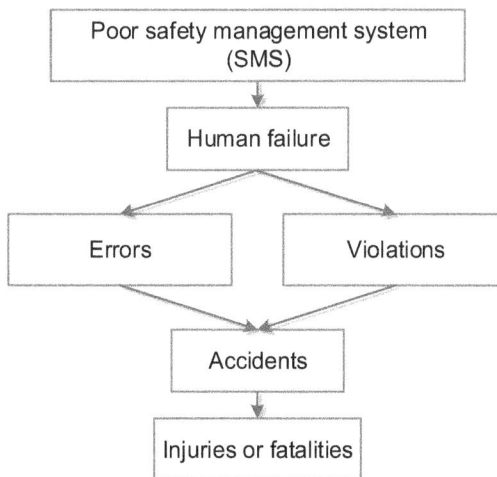

Figure 3.1 Classification of poor SMS causing accidents

safety management is used to oversee the safety regulations, standards, and procedures on construction sites. In addition to safety regulations, safety guidelines significantly contribute to safety management (Khalid et al., 2021). This chapter provides in-depth data on how the integration of SMS could reduce human failure on construction sites. The sections that follow in this chapter address the impact of construction SMS. The research method is outlined, followed by the findings and discussions.

3.2   SMS in construction

SMS is a system created to simplify managing safety and reduce industry accidents (Yiu et al., 2019). It is described as a managerial procedure that may involve several instruments, whether or not they are required, to enhance safety performance (Bridi et al., 2021). It enables managers' active involvement in its implementation to provide opportunities for workers to become familiar with HSW-related policies and procedures. This implies that safety procedures would command high regard when standards are compared. However, the adoption and modification of safety standards and the implementation of SMS remain significant challenges despite the vast research that has been conducted (Wu et al., 2021).

According to Accou and Reniers (2019), an SMS is developed to eliminate hazards to prevent accidents in the workplace. However, despite the adoption of SMS in construction, the industry continues to record many accidents (Fan et al., 2014). Because it is assumed that accidents occur when workers deviate from the required procedures, solutions centre on putting even more pressure on workers to comply (Provan et al., 2020). Accidents in the construction industry negatively impact the economy and society because they frequently result in fatalities and severe injuries. Common HSW hazards are identified through accident analysis and added to the body of information that will enable future SMS improvements (Xu et al., 2021). However, all HSW initiatives depend mostly on how well the SMS is implemented (Lee et al., 2020). One advantage of adopting a formal management system to implement SMSs is using risk profiling to establish an organisation's operational H&S objectives (Chi & Lin, 2022). The basis for identifying safety hazards must be the proper classification, evaluation, and identification procedure for significant issues. In addition, a proper method for classifying, analysing, and diagnosing important issues from a safety database must be used to identify safety hazards. Furthermore, SMSs are frequently institutionalised with safety components such as policies, goals, objectives, processes, responsibilities, and other safety measures that address known or foreseeably dangerous situations. As it frequently assumes that safety issues are caused by deviations that need to be eliminated, the SMS strategy concentrates on minimising uncertainty rather than managing it (Pilanawithana et al., 2022). As a result, SMSs are comprised of several components. Some of the components of the SMS that could be utilised are presented in Table 3.1.

Table 3.1 Components of the safety management system (SMS)

Code	Components of the SMS	Sources
HI	Hazard identification	Chi and Lin (2022); Pilanawithana et al. (2022); McKinnon (2019)
SI	Safety inspection	Chi and Lin (2022); Yiu et al. (2019)
OHST	Occupational health, and safety (OHS) training	Chi and Lin (2022); Lee et al. (2020); Yiu et al. (2019)
SWP	Safe work procedures	Pilanawithana et al. (2022); Alli (2008)
RSP	Risk assessment plan	Pilanawithana et al. (2022); McKinnon (2019)
SRS	Safety reporting system	McKinnon (2019)
TT	Toolbox talk	Alli (2008)
OHSP	OHS policy	Chi and Lin (2022); Pilanawithana et al. (2022); McKinnon (2019); Yiu et al. (2019)

3.3 Research methods

This chapter shows how implementing an SMS could mitigate human failure on construction sites. A mixed-methods approach was adopted to address the phenomenon. According to Plano Clark and Ivankova (2016), mixed methods is a coordinated process for collecting data using qualitative and quantitative techniques. Case study research was used to collect the qualitative data, while survey questionnaires were used to collect the quantitative data. The case study research strategy involved multiple case studies in the South African city of Bloemfontein. Multiple case studies were used in this study to facilitate the investigation into how SMSs may be implemented to reduce human failure on construction projects. It was decided to use building projects as case studies because of the nature of construction work, which exposes workers to risks that could result in accidents if they are not avoided. Purposive sampling was utilised in the case study selection method because the study placed great emphasis on construction projects. The construction manager's consent to utilised their project as a case study was first requested. Interviews were subsequently conducted with construction managers, site engineers, foremen, artisans, and labourers for the multiple case study project (Table 3.2).

The data was collected in the multiple case studies using open-ended questions in semi-structured and focus group interviews (Flick, 2014). The study's main goals were explained to the interviewees before the semi-structured interviews commenced in the multiple case studies. This helped the interviewees understand the reported SMS issue under investigation. In addition, focus groups were conducted with workers and artisans over lunch. Due to the adoption of focus group interviews, every interviewee had the opportunity to respond to the questions without being interrupted by the other interviewees. At the project site where they were working, they were also able to discuss the issues concerning the safety system

Table 3.2 Multiple case studies sample

Case study code	Case-study projects	Interviewees code	Members of the construction teams (interviewees)	Number
BP1	Building project 1	BP1-CM	Construction manager	1
		BP1-SM	Safety manager	1
		BP1-SE	Site engineer	1
		BP1-SF	Senior foreman	1
		BP1-JF	Junior foreman	1
		BP1-GA	Group of artisans	4
		BP1-GW	Group of workers	9
BP2	Building project 2	BP2-SM	Safety manager	1
		BP2-CM	Construction manager	1
		BP2-SSE	Senior site engineer	1
		BP2-SSQS	Senior site quantity surveyor	1
		BP2-SF	Senior foreman	1
		BP2-JSE	Junior site engineer	1
		BP2-JF	Junior foreman	1
		BP2-SS	Student supervisor	
BP3	Building project 3	BP3-CM	Construction manager	1
		BP3-SM	Safety manager	1
		BP3-SE	Site engineer	1
		BP3-SF	Senior foreman	1
		BP3-JF	Junior foreman	1
		BP3-GA	Group of artisans	7
Total interviewees				**39**

adopted on their project site. In BP1, the focus group interviews involved a group of four artisans who were bricklayers, and two groups of workers were divided into two groups. The first group of workers comprised five workers, and the remaining four workers made up the second group while in BP3, the focus group interview involved a group of seven artisans who were bricklayers. The focus group interviews in BP1 and BP2 were conducted during the lunch break.

As a result of the multiple case studies, the researchers conducted in-person interviews that lasted between 30 and 60 minutes each between July 2018 and April 2019. The decision was made to record each interview with the interviewees' permission. These recordings were later transcribed. A thematic analysis was then used to analyse the qualitative data (Bryman, 2012).

Concerning the quantitative data, a survey questionnaire was adopted per the recommendation of Clark and Ivankova (2016). Survey questionnaires were used to supplement the qualitative data collected in the multiple case studies. The eight SMS components were the subject of the survey questionnaires, which were based on the literature review shown in Table 3.1. This was due to the hypothesis that the eight identified components of the SMS could lead to human failure if inadequately addressed on construction sites.

The quantitative data was collected using an online survey tool. The survey tool was designed, and the link to the survey was later sent to the South African

Table 3.3 Participants' demographic profile

Participants	Number	Percentage
Contractors	78	48.00
Clients	27	17.00
Members of the consultants' civil engineering companies	8	5.00
Members of the consultants' health and safety companies	40	25.00
Members of the consultants' architect companies	5	3.00
Members of the consultants' quantity surveying companies	4	2.00
Total participants	**162**	**100.0**

Council for Project and Construction Management Professionals (SACPCMP). The SACPCMP distributed the link to its members to complete the survey. The survey was sent to SACPCMP members registered as professional safety agents, safety managers, and construction managers. In addition, the link was sent to candidate safety officers. As a result, it can be concluded that random sampling was used to select the participants (Plano Clark & Ivankova, 2016). The demographic profile of the participants is shown in Table 3.3.

A Likert scale of 0–5 was adopted to design the survey questionnaires. A Likert scale is summurised as follows: 0 = Unsure, 1 = Minor, 2 = Above minor, 3 = Neutral, 4 = Near major, and 5 = Major. After collecting the quantitative data, the data were analysed through the Statistical Package for Social Science (SPSS). In addition, the descriptive data were analysed using a mean score (MS) and standard deviation (SD). As a result, the MS helped to measure the average score of the statistical data while the SD measured the variability. Cronbach's α was used to measure the reliability of the statistical data (Plano Clark & Ivankova, 2016). The quantitative data was collected between 15 July and 15 October 2019.

3.4 Results

3.4.1 *SMS reducing human failure in construction sites*

This section presents the analysed qualitative data collected through multiple case studies. It was found that the multiple case studies presented similar results. Only people who were working on the multiple case projects were interviewed. This was because people needed to respond to questions concerning the construction projects they worked on.

Regarding contractors implementing SMSs on their construction projects, most of the interviewees in the multiple case studies responded that their employers (contractors) had employed a safety manager responsible to design the SMS to monitor the H&S of the construction projects. In the South African construction industry, the SMS is known as safety files designed by the safety manager before the project starts and approved by the clients. Moreover, BP2-SM in BP2 explained that their contractor had created a safety management team led by a safety manager to design effective safety files according to the Construction Regulations 2014. This system is responsible for managing the H&S of PiC to prevent accidents and illness. In addition, the safety files are categorised into several components, including a hazard identification section, a risk assessment tool section, a safe work procedures section, a safety training and observation section, and a safety inspections section. The BP2 is quoted as follows:

> *Since the start of the project, we have not recorded accidents; this is because of the team is managing the H&S of our people according to the designed safety files. As a team, we did not just design the safety files for the client to approve, but we designed the system for our people to implement.*

A BP1-CM in BP1 corroborated the statement by BP2-SM in BP2 that their employer created a safety management team at the start of the project, and their first task was to design safety files that were then sent to the client to approve. The safety files were designed to assist the company's compliance with the designed safety rules and regulations. The safety managers are responsible for managing the system to prevent accidents and promote a safe working environment for construction workers.

However, the BP1-GW in BP1 had different responses from those of BP1-CM in BP1 and BP2-SM in BP2 concerning how the SMS is adopted in the construction industry. This is because the BP1-GW in BP1 indicated that, despite the design of the safety file, the workers continued to be exposed to safety risks that often result in accidents. The BP1-GW in BP1 is quoted as follows:

> *The implementation of safety files is critical to promote a safe work environment, but our foremen tend to instruct us to ignore the safety procedure because they are often under pressure to finish tasks in time. As a result, their actions often promote an unsafe work environment because we are instructed to ignore certain safety procedures.*

Additionally, the BP3-GA in BP3 highlighted that the safety system is designed to help workers understand the company's OHS rules and regulations and to educate them on how to work safely, particularly when using tools and equipment that frequently result in hand injuries. They indicated that at the beginning of the project, they attended safety training, where they learned how to operate machines and how crucial it is to operate machines safely.

The interviewees in BP2 noted that, in contrast to BP3, where they received training on using machines properly, they needed help with machine operations in their project. This was due to BP2's report that a worker accidentally had his finger amputated by a concrete mixer drum. According to reports, a worker was pouring concrete while standing next to a concrete drum when he lost his balance and got his hand caught in the drum, losing a finger.

Regarding the procedure that must be followed when developing SMSs on construction sites, most interviewees in the multiple case studies stated that they developed their safety system in compliance with the Construction Regulations 2014. The Construction Regulation 2014 is administered by the Department of Labour in South Africa. The safety management team complies with the Occupational Health and Safety Act 85, 1993, the Construction Regulations 2014. The designed safety system includes elements such as the organisational organogram, accountability, and responsibility for health and safety, a method statement, hazard identification, risk assessment, safety training, toolbox talk, and safety inspection records. Some of the interviewees' responses were as follows:

> *As the project leaders, we are accountable for developing a system that would protect everyone working on our project. The Construction Regulations of 2014 is used as a reference guide to design the safety files. The safety system identifies risks and eliminates them before they could cause accidents (BP3-SM and BP3-CM in BP2).*
>
> *The safety management team led by a safety manager is responsible for investigating and studying the safety guidelines provided by the client. This would assist in designing appropriate safety files that the client would approve if it met the safety standards stipulated in the Construction Regulation 2014 (BP1-SM).*

In addition, the BP1-CM and BP2-SS explained that it is important that the safety management team should design the safety files according to the client's recommendation. This is because failure by the client to approve the safety files would mean that the contractor was not permitted to commence with the construction. Therefore, the safety files must be designed correctly as their approval is the first permit for the contractor to establish the site. BP3-SF and BP3-JF further reported that according to the safety file, they are responsible for installing the safety warning signage. Safety warning signage must be installed on the walkway, project entrance, excavated area, drive through, and hazardous areas. Safety warning signage must be designed according to safety standards, and written clearly and correctly in a language everyone on site understands.

Concerning the impact of human failure on construction sites, it was reported by several interviewees from the multiple case studies that the adoption of a risk assessment tool assists in identifying and analysing risks that have the potential to cause human failure. For example, BP1-SM, BP3-SE, and BP3-SF had a similar response, namely that the causes of human failure were rooted in the unsafe behaviour of workers on site. Workers get used to the working process, ignoring the safety protocols because they become complacent, having never experienced any incidents; however, such behaviour might lead to accidents. Moreover, BP2-JSE noted that the desire by managers to achieve higher levels of productivity often compels workers to ignore safety regulations. He further explained that one of the reasons they have poor housekeeping was because their manager needed to be more focused on completing the task on time and paid less attention to cleaning up the project area.

In addition, BP1-SF and BP2-SE had similar responses regarding the impact of poor housekeeping on both projects. The interviewees explained that in both projects, they struggled to keep the project site clean; the poor housekeeping was caused by the people working in construction. Also, poor housekeeping could result in accidents connected to slips and fall as wasted materials block the walkways. This is an example of human error and violations. The BP1-SF further explained that on his project site, they struggled to keep the project area clean because they needed to catch up and could not afford to task workers to clean up the entire project. In addition, BP2-SS explained that everyone on his project site knows the importance of good housekeeping; nevertheless, poor housekeeping continues on their site, as broken bricks are scattered all over the project.

Regarding the impact of SMS towards hazard and risk control, it was reported by most interviewees in the multiple case studies that their safety management teams had identified the potential hazards and risks for their tasks. In addition, during the toolbox talk, they presented the identified hazards according to their activities and guided workers on how to minimise and prevent risks from causing accidents. However, despite identifying the hazards and assessing the risks, the projects continued to experience accidents. For example, it was reported that in BP2, a worker's hands were ripped off by a ready-mix concrete drum. In addition, BP3-SM reported that a worker fell from the roof truss while cutting the rafters on their project. Such cases raise a question concerning the implementation of the safety system, as accidents happen despite the safety managers having identified hazards and discussed the risk of such hazards with the workers.

It was reported that in BP3, the workers were wearing safety harnesses to prevent falling, and the injured worker was also wearing a safety harness but failed to attach it to the roof truss. It can be concluded that safety compliance by the workers and safety monitoring by management are not enforced. The blame for such accidents should be placed at the worker's door for failing to install the safety harness on the roof truss and the management for failing to stop the workers from cutting the rafters without attaching the safety harness. In addition, BP2-SSQS highlighted that such accidents should also be blamed on the type of safety training that is conducted on-site. This is because there was no facility that the safety management team could use to conduct safety training; in most cases, safety training was introduced briefly and presented in detail using practical examples.

3.4.2   *The assessment of the components of the SMS*

This section presents the quantitative data that was aimed at evaluating the identified eight components of the SMS that have the potential to cause human failure on construction sites. The assessment of the components of the SMS was rated using a Likert scale of 0 (Unsure) to 5 (Major). The impact of the components of the SMS towards human failure on construction sites are shown in Table 3.4. It should be noted that HI is the highest ranked at an MS 3.80 while OHSP is ranked the lowest at an MS 2.81. SI is ranked second at an MS 3.61, OHST is ranked third at an MS 3.54, SWP is ranked fourth at an MS 3.44, RAP is ranked fifth at an MS 3.42,

Table 3.4 Components of the SMS on construction sites

SMS components	MS	SD	Rate
HI	3.80	1.434	1
SI	3.61	1.453	2
OHST	3.54	1.483	3
SWP	3.44	1.448	4
RAP	3.42	1.486	5
SRP	3.41	1.493	6
TT	3.01	1.548	7
OHSP	2.81	1.498	8
Cronbach's α			**0.993**

SRP is ranked sixth at an MS 3.41 and TT is ranked seventh at an MS 3.01. It is observed that seven of the eight components of the SMS ranked above 3.00. This is an indication that these SMS components are significant and should be addressed to reduce human failure on construction sites. This statistical data corroborates the results from the case study concerning their experienced accidents rooted in poor SMS.

It is also observed in Table 3.4 that the reliability test of the components of the SMS was measured at 0.993. Therefore, it is safe to say that this reliability test is acceptable because it is above the minimum of 0.70.

3.5 Discussion

This study investigated how the implementation of the SMS would mitigate human failure on construction sites. The multiple case studies show that contractors design an SMS, known as safety files, which are approved by the clients before the start of the project. This highlights the value of SMS in the construction industry, because it is a requirement that is evaluated by the client. Because Safety-I aims to align and regulate the company and its workforce through the deployment of SMS, the implementation of an SMS does in fact relate to it. The SMSs in South Africa are designed according to the Construction Regulations 2014. For example, BP3_SM explained that he designed a safety system that would protect PiC. He mentioned that the design referred to the Construction Regulation 2014. This practice aligns with the notion that an SMS is developed to simplify managing safety and reduce workplace accidents (Yiu et al., 2019). As a result, SMSs are aligned with Safety-I and Safety-II.

The safety management literature described how SMSs are frequently institutionalised with policies, goals, objectives, processes, responsibilities, and other precautionary measures that address known dangerous situations. For instance, BP2 reported that a ready-mix concrete drum ripped off the hands of a worker, perhaps owing to poor SMS as the event was linked to an error. Another example is from BP3 who reported that a worker fell from the roof while cutting the rafters. This accident is linked to both errors and violations because such action was unintended (error) and non-compliant in terms of not using a safety harness (violation) while working at height. These two cited accidents are indicative of poor SMS.

It can, therefore, be concluded that human failure in the form of errors and violations experienced in BP2 and BP3 are the result of poor SMSs that relate to Safety-I and Safety-II. To prevent such accidents, it is important to integrate the components of the SMS on construction sites. Table 3.4 shows that the components of the SMS are ranked above the midpoint of 3.00, which suggests that they could be clearly defined and implemented. For example, the BP3-SM should have prevented the worker from falling if the risk had been identified, and the provision of a safety harness had been made available to the workers. This thus explains why McKinnon (2019) stated that the components of an SMS all work together to simplify workplace recognition and management. Therefore, these components of the SMS should be integrated; this would help to prevent the occurrence of human failure and improve SMS on construction sites.

Figure 3.2 shows the implementation of the SMS components to mitigate human failure. In the integrated model, the components of the SMS are connected to each

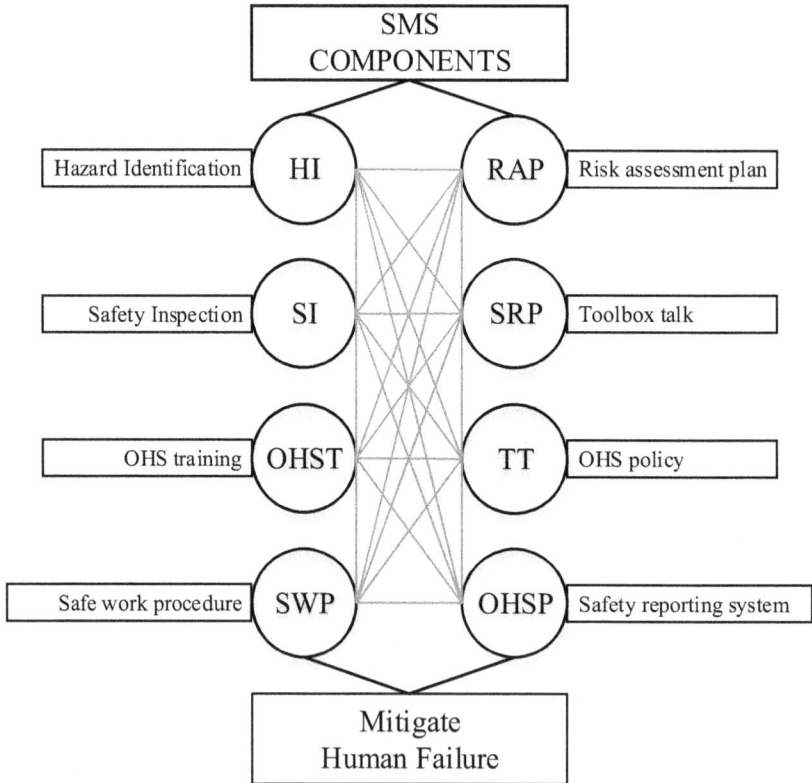

Figure 3.2 Integrated SMS model to mitigate human failure

other. The integration of each component would help to improve the SMS and ensure that it is implemented seamlessly to prevent human failure. The prevention of human failure would promote an effective SMS that would reduce accidents on construction sites.

3.6   Chapter summary

SMS aims to create a system that could manage safety and reduce accidents in the industry. Previous studies and the one reported in this chapter attest that accidents are curtailed when SMS is effective. However, poor SMS often results in accidents frequently occur on construction sites. Human failure in the form of errors and violations also occurs regularly on construction sites. Figure 3.2 indicates the value of considering project teams in construction operations. The integrated SMS model could prevent human failure if the components are not stand-alone.

References

Accou B and Reniers G (2019) Developing a method to improve safety management systems based on accident investigations: The safety fractal analysis. *Safety Science* 115: 285–293.

Alli BO (2008) *Fundamental principles of occupational health and safety*, 2nd ed. Geneva: International Labour Organization (ILO).

Bridi ME, Formoso CT and Saurin TA (2021) A systems thinking based method for assessing safety management best practices in construction. *Safety Science* 141: 105345. Available from: https://doi.org/10.1016/j.ssci.2021.105345

Bryman A (2012) *Social research methods*, 4th ed. Oxford: Oxford University Press.

Chi C-F and Lin Y-C (2022) The development of a safety management system (SMS) framework based on root cause analysis of disabling accidents. *International Journal of Industrial Ergonomics* 92: 103351. Available from: https://doi.org/10.1016/j.ergon.2022.103351

Fan D, Lo CK, Ching V, et al. (2014) Occupational health and safety issues in operations management: A systematic and citation network analysis review. *International Journal of Production and Economics* 158: 334–344.

Federated Employers Mutual Assurance Company (FEM) (2022, April 26) *FEM registers alarming number of injuries in local construction industry*. Available from: https://zeroisnoaccident.co.za/fem-registers-alarming-number-of-injuries-in-local-construction-industry/

Flick U (2014) *An introduction to qualitative research*, 5th ed. London: Sage Publishing.

Ghasemi F, Gholamizadeh K, Farjadnia A, et al. (2022) Human and organizational failures analysis in process industries using FBN-HFACS model: Learning from a toxic gas leakage accident. *Journal of Loss Prevention in the Process Industries* 78: 104823. Available from: https://doi.org/10.1016/j.jlp.2022.104823

Health and Safety Executive (HSE) (2009) *Reducing error and influencing behaviour*. London: Health and Safety.

Health and Safety Executive (HSE) (2019) *Kinds of accident statistics in Great Britain*. London: The Health and Safety Executive. Available from: www.hse.gov.uk/statistics/causinj/kinds-of-accident.pdf

Hollnagel E (2018) *Safety-I and safety-II: The past and future of safety management*. Boca Raton: CRC Press.

Khalid U, Sagoo A and Benachir M (2021) Safety management system (SMS) framework development – Mitigating the critical safety factors affecting health and safety performance in construction projects. *Safety Science* 143: 105402. Available from: https://doi.org/10.1016/j.ssci.2021.105402

Koc K and Gurgun AP (2022) Scenario-based automated data preprocessing to predict severity of construction accidents. *Automation in Construction* 140: 104351. Available from: https://doi.org/10.1016/j.autcon.2022.104351

Lee Y-C, Shariatfar M, Rashidi A, et al. (2020) Evidence-driven sound detection for prenotification and identification of construction safety hazards and accidents. *Automation in Construction* 113: 103127. Available from: https://doi.org/10.1016/j.autcon.2020.103127

Martins JB, Carim G, Saurin TA, et al. (2022) Integrating safety-I and safety-II: Learning from failure and success in construction sites. *Safety Science* 148: 105672. Available from: https://doi.org/10.1016/j.ssci.2022.105672

McKinnon RC (2019) *The design, implementation, and audit of occupational health and safety management systems*. New York: CRC Press.

Pilanawithana NM, Feng Y, London K, et al. (2022) Developing resilience for safety management systems in building repair and maintenance: A conceptual model. *Safety Science* 152: 105768. Available from: https://doi.org/10.1016/j.ssci.2022.105768

Plano Clark V and Ivankova NV (2016) *Mixed method research: A guide to the field*. Los Angeles, CA: Sage Publications.

Provan DJ, Woods DD, Dekke SW, et al. (2020) Safety II professionals: How resilience engineering can transform safety practice. *Reliability Engineering & System Safety* 195: 106740. Available from: https://doi.org/10.1016/j.ress.2019.106740

Ranasinghe U, Jefferies M, Davis P, et al. (2020) Resilience engineering indicators and safety management: A systematic review. *Safety and Health at Work* 11(2): 127–135. Available from: https://doi.org/10.1016/j.shaw.2020.03.009

Reason J (2000) Human error: Models and management. *British Medical Journal* 320: 768–770.

Reason J (2016) *Organizational accidents revisited*. New York: CRC Press.

Woodward S (2019) Moving towards a safety II approach. *Journal of Patient Safety and Risk* 24(3): 96–99. Available from: http://doi.org/10.1177/2516043519855264

Wu H, Zhong B, Li H, et al. (2021) Combining computer vision with semantic reasoning for on-site safety management in construction. *Journal of Building Engineering* 42: 103036. Available from: https://doi.org/10.1016/j.jobe.2021.103036

Xia J, Liu Y, Zhao D, et al. (2021) Human factors analysis of China's confined space operation accidents from 2008 to 2018. *Journal of Loss Prevention in the Process Industries* 71: 104480. Available from: https://doi.org/10.1016/j.jlp.2021.104480

Xu N, Ma L, Liu Q, et al. (2021) An improved text mining approach to extract safety risk factors from construction accident reports. *Safety Science*: 105216. Available from: https://doi.org/10.1016/j.ssci.2021.105216

Yiu NSN, Chan DWM, Shan M and Sze NN (2019) Implementation of safety management system in managing construction projects: Benefits and obstacles. *Safety Science* 117: 23–32. Available from: http://doi.org/10.1016/j.ssci.2019.03.027

4 Minimising human errors with learning-by-doing

4.1 Introduction

Construction is a dangerous, demanding, and stressful industry that can seriously affect workers' occupational health and safety (OHS). Construction is considered one of the highest-risk industries worldwide (Fang et al., 2022). For instance, in South Africa, between 2015 and 2021, there were 5,842 'falling' cases and 18,165 'struck by' cases, accounting for 44% of all accidents and 35% of fatal accidents (Federated Employers Mutual Assurance Company [FEM], 2022). Additionally, it has been reported that between 2011 and 2020, 9,625 construction workers died in accidents at work in the United States of America (USA; CPWR, 2022). Another example of this risk is falling from heights, which accounts for 36% of fatal injuries in the USA (CPWR, 2022). Falling from heights is also the leading cause of fatalities in the Australian construction industry (Jonathan et al., 2022).

The high accident rate in the construction industry, as illustrated in the preceding paragraph, is a major concern, despite efforts to reverse the trend. To develop preventive measures, a further understanding of the factors contributing to construction accidents is needed (Cabello et al., 2021). In the construction industry, four major categories of accidents might occur: falls, electrocutions, caught-in between, and struck-by (Li et al., 2023). Human error is one of the primary factors leading to accidents in construction projects (Rafieyan et al., 2022). Such human error that results in accidents is caused by attitudes and behaviours (slip-ups and lapses), perceptions and decision-making (cognitive biases or heuristics), and actions (Reason, 2000). It is important to recognise that human error results from the organisations and environments in which people work on construction projects (Reason, 2016). The underlying conditions for human error to manifest in construction projects are frequently created by the strategic decisions made by managers or decision-makers (Love, 2021).

Several studies have investigated various methods to reduce the occurrence of human error and improve site safety. Some studies, for instance, concentrated on the execution of a human risk factor management plan, ongoing training for both seasoned and new personnel, the efficient use of fall protection equipment, and supervisors' commitment to safe behaviour (Jonathan et al., 2022). While some studies address the industry's human error problem by concentrating solely

DOI: 10.1201/9781003365341-4

on organisational and physical risks, little focus has been placed on adopting learning-by-doing to improve site safety. Thus, learning-by-doing is distinguished from learning through observation of others performing, reading instructions or descriptions, or talking lectures from others (Reese, 2011). Because learning can only occur in the workplace without making errors, learning by doing is recommended. Therefore, by creating, trying, comprehending, and altering a strategy, construction practitioners can practise persistent problem-solving with the help of learning-by-doing.

4.2   Learning-by-doing in industrial environments

Many construction tasks still demand human discretion, adaptability, and agility, necessitating ongoing learning. As a result, learning-by-doing offers the construction industry an opportunity to teach workers how to reduce human error and improve site safety. Bruce and Bloch (2012) defined learning-by-doing as the finest method for helping individuals (construction workers) make sense of their experiences, especially when they actively take part in creating things and discovering new places. Learning-by-doing is a philosophical postulate that can be applied to various learning scenarios, particularly in the construction industry. Some would even argue that it applies to all learning. It is also a pedagogical strategy that teachers or managers use to encourage their students or workers to learn in more engaging, novel ways. Such a learning process would improve site safety and reduce human error, which is a major contributor to accidents in the construction industry.

Reese (2011) identified the learning-by-doing principle as trial-and-error learning or discovery versus instruction, practical experience versus book learning, the practice-theory-practice dialectic, and proof upon practice. These principles promote practical learning, which is incorporated into construction since the industry is labour intensive (Fang et al., 2020). As a result, most construction project activities are completed by people using their hands and learning the required project activities. The fundamental principle of learning-by-doing is emphasised by hands-on learning opportunities (Mohd et al., 2019). As part of the learning-by-doing process, people working in construction, for example, can practise persistent problem-solving by developing, experimenting with, interpreting, and modifying a playing strategy. This is because the application of learning-by-doing promotes the most meaningful learning experiences that might be relevant to the construction industry. Learning by experience is promoted in several learning theories, including Dale's cone of learning and the learning pyramid (Ahmad et al., 2018). In addition, dynamic systems can be used to teach learning by doing. It would enable learners to see and practice the underlying concepts, helping them improve their organisational and systemic thinking skills.

Thus, this chapter proposes learning-by-doing to solve human error causing construction accidents, as Rafieyan et al. (2022) reported. Human error is a significant source of knowledge because it reveals when a process goes wrong. Employees can learn from errors by being aware of the aspects that affect learning from errors. Doytchev and Hibberd (2009) contended that learning from errors and occurrences is essential for organisations to become safer and for their operational procedures to function more effectively. Learning from errors is one method to make construction sites safer. To what extent errors are discussed in the field may influence how much learning from errors occurs (Horvath et al., 2021). Errors are a warning sign that something is wrong and that action is needed to prevent accidents.

Learning in the construction industry should be encouraged because general workers are usually under-educated, lack safety knowledge, and must be made aware of the risks of irregular operations (Xu et al., 2019). Therefore, integrating learning-by-doing with safety training may help reduce the risk of human error in construction projects. Integrating safety training and learning-by-doing may lead to a better understanding of error characteristics and how the cultural environment impacts learning from errors. This can help develop effective interventions to improve learning from errors (Horvath et al., 2021). Therefore, rather than depending solely on the transmission of knowledge, learning-by-doing in the construction industry would assist workers in comprehending how to complete their tasks safely. This is because learning-by-doing, a type of work-based learning, relies significantly on accruing work experience to increase effectiveness (Ye et al., 2019).

4.3 Research methods

To answer the chapter research question, namely 'How can the use of learning-by-doing mitigate human errors on construction sites?', the study used mixed-methods research. The research choice permitted the integration of both qualitative and quantitative data (Plano Clark & Ivankova, 2016). The qualitative data were collected using case studies (Hancock & Algozzine, 2011). Three different building works projects in Bloemfontein, South Africa, served as the basis for the case studies, including multiple case studies highlighted in Table 4.1. Buildings were chosen as the type of construction project because they expose workers to accidents such as falls, electrocutions, being stuck between objects, and being struck, as Li et al. (2023) reported. Purposive sampling was used to select interviewees because of the need to engage people in construction who were employed on the projects.

To collect the qualitative data for the multiple case studies, open-ended questions through semi-structured and focus group interviews were used. The study's goal, namely to determine whether using learning-by-doing would reduce human errors on construction sites, was discussed with the respondents. This made it easier for them to comprehend the concept of human error and how learning-by-doing could address it. Each interviewee was permitted to discuss their real-world experience during the interview. In addition, an open interview discussion was adopted during the focus group interviews. Table 4.1 shows two focus groups for Building Project 1 (BP1), one with four artisans and the other with five and four workers. The Building Project 3 (BP3) focus group consisted of seven artisans. The interviews for the multiple case studies were conducted at lunch time because of the participants' busy schedules.

The quantitative data were collected through an online questionnaire (Plano Clark & Ivankova, 2016). The questionnaire survey was designed

Table 4.1 Multiple case study project details

Case studies	Code	Interviewees	Code	Number
Building project 1	BP1	Construction manager	BP1-CM	1
		Safety manager	BP1-SM	1
		Site engineer	BP1-SE	1
		Senior foreman	BP1-SF	1
		Junior foreman	BP1-JF	1
		Group of artisans	BP1-GA	4
		Group of workers	BP1-GW	9
Building project 2	BP2	Safety manager	BP2-SM	1
		Construction manager	BP2-CM	1
		Senior site engineer	BP2-SSE	1
		Senior site quantity surveyor	BP2-SSQS	1
		Senior foreman	BP2-SF	1
		Junior site engineer	BP2-JSE	1
		Junior foreman	BP2-JF	1
		Student supervisor	BP2-SS	1
Building project 3	BP3	Construction manager	BP3-CM	1
		Safety manager	BP3-SM	1
		Site engineer	BP3-SE	1
		Senior foreman	BP3-SF	1
		Junior foreman	BP3-JF	1
		Group of artisans	BP3-GA	7
Total interviewees				**39**

to evaluate the learning-by-doing factors. Table 4.3 shows the identified 15 learning-by-doing factors categorised into worker- and management-induced factors. The purpose of evaluating these learning-by-doing-induced factors was to validate the qualitative data collected through multiple case studies. The survey questionnaire was designed using a Likert scale of 0–5, where 0 = Unsure, 1 = represented Minor, 2 = represented Above minor, 3 = represented Neutral, 4 = represented Near major, and 5 = represented Major. The South African Council for Project and Construction Management Professionals (SACPCMP) shared the self-administered questionnaire URL with its members through its database. As a result, a simple random sample technique was used to choose the participants (Plano Clark & Ivankova, 2016). The study used the Statistical Package for Social Science (SPSS) to analyse the quantitative data. A mean score (MS) and standard deviation (SD) were used to analyse the descriptive data. To determine the average score of the statistical data, the authors utilised the MS, while the SD measured the variability of the MS. The Cronbach's α was used to measure the reliability of the statistical data (Plano Clark & Ivankova, 2016). The quantitative data was collected between 15 July and 15 October 2019. Table 4.2 shows the participants' demographic profile.

Table 4.2 Participants' demographic profile

Participants	Number	Percentage
Contractors	78	48.00
Clients	27	17.00
Members of the consultants' civil engineering companies	8	5.00
Members of the consultants' health and safety companies	40	25.00
Members of the consultants' architect companies	5	3.00
Members of the consultants' quantity surveying companies	4	2.00
Total participants	**162**	**100.0**

4.4 Results

This section contains two sub-sections that summarise the research findings. The qualitative data is presented in the first sub-section, while the quantitative data is included in the second sub-section as a discussion.

4.4.1 Learning-by-doing and human errors

As a result of the widespread use of learning-by-doing on construction sites, some interviewees from multiple case studies said that the use of learning-by-doing in construction was carried out in an informal training program to transfer work knowledge and skills without referencing any theoretical concept or structural learning program. The literature also referenced that learning-by-doing is a peda-gogical strategy that managers use to encourage their workers to learn in more engaging, novel ways. For example, a BP2-CM is quoted as follows:

> *It should be acknowledged that, on construction sites, a learning or skills transfer does not occur in a classroom or a boardroom. Our workers are taught or learn how to carry out the works by observing the skilled work-ers known as artisans while completing their task and listening to their instructions.*

In addition, some of the interviewees in BP1 described the process of learning-by-doing on their construction projects. They explained that learning-by-doing is a process in which artisans transfer work skills to inexperienced workers. The inex-perienced workers had to develop a culture of being observant and asking ques-tions all the time because the skilled workers needed more time to teach them formally about how to do the work. Figure 4.1 shows an observed example of the workers casting a concrete footing in BP1. Additionally, it was noted that the work-ers were using the appropriate personal protection equipment (PPE) and concrete casting equipment. This shows that learning involves changing one's knowledge, abilities, attitudes, and behaviour significantly.

Figure 4.1 Workers casting a concrete footing

It is further reported that BP3-SM and BP3-CM suggested that the learning-by-doing should be integrated with the safety training program designed on construction projects. This is based on the notion that safety training is conducted at the beginning of the project and that integrating it with learning-by-doing would enable the workers to learn about the safety regulations while they are carrying out their tasks. This would improve their understanding of the safety procedures. Learning-by-doing would also improve workers' motivation for safety, allowing them to learn from their errors and preventing future accidents. In addition, BP3-SM reported that they are responsible for designing the safety system, therefore, it would be beneficial in the future to incorporate learning-by-doing as it would help them to introduce practical safety training on construction sites.

With respect to the impact of knowledge and skills transfer on project sites, it was reported by a BP1-CM that on their sites, the foremen were responsible for transferring the skills of workers to their new recruits, especially students who were completing their in-service (practical) training. At the same time, the artisans were responsible for transferring the work skills to the general workers. Because the workers are taught construction work first-hand, this type of learning may be related to learning-by-doing. Additionally, this method would assist the workers in understanding the construction activities because every error they make helps develop their work. It is well known that any worker that commits an error may affect how it is addressed and that this experience may inspire their learning and alter their perspective. People are said to learn more from their errors; therefore, learning-by-doing would increase workers' understanding of their work. Therefore, BP1-CM is quoted as follows:

I always tell my new workers, especially the recently recruited student engineers or construction managers, that if they want to learn the real construction work, they must behave like workers and follow the instruction of the foremen. He further explained that the foremen would teach them how to solve practical problems, and they would also learn the behavioural acts of the workers, which would be helpful for them to know how to manage the workers.

BP2-SM reported that when he was still doing his practical work, he preferred to work with his foreman, from whom he learned the importance of having an appropriate safety system and how to prevent hazards that might result in accidents, without disrupting the performance of the workers. In addition, BP3-JSE and BP3-SS stated that some of the workers and artisans working on their construction project were often taken to attend short courses that included scaffold erecting, first-aid training, and carpentry. The interviewees explained that most of their skilled workers, who were promoted to become foremen, were taken on a short course to learn how to erect a scaffold, especially because most resident engineers require someone who is certified to instal scaffolding on construction projects. This is because scaffolding is identified as a hazard that often results in accidents because of poor installation or the use of damaged props when installing the scaffolds. Accidents involving scaffolding frequently involve falls from heights that happen as a result of human error. It is crucial to understand that when a scaffold collapses, it is usually because of human error. Thus, it is essential that learning is introduced in the right way to prevent such errors and help workers learn from their past errors to prevent such accidents from happening again. In addition, Figure 4.2 shows a scaffold inspection notification observed in BP3.

4.4.2   *Learning about the root causes of accidents*

Furthermore, it was reported by most of the interviewees that in terms of the reduction of human error, learning about the root causes of the accidents would help to understand the actual factors causing a human error. This is because errors in construction can be made by anyone, regardless of their knowledge level. In addition, errors can occur in construction for various reasons, including fatigue, stress, behaviour, and unforeseen conditions. Thus, BP2-CM explained that learning about the causes of human error would help in understanding the behaviour or actions of the workers and the reason why they often ignore the safety regulations. Furthermore, according to BP3-SE, understanding human error would enable workers to recognise risks and offer solutions for reducing them to prevent accidents. For instance, recognising fall risks would assist in preventing errors that could be brought on by workers' unsafe behaviour when working at heights. It is

Figure 4.2 Scaffold inspection notice

widely known that when workers do not follow safety rules, they frequently engage in unsafe behaviour that could lead to falls when working at heights.

Additionally, BP1-SF and BP3-SSE stated that they have observed from experience that workers frequently contribute to errors because of how the construction projects are designed. They noted that most clients require unique projects, which become complex and often expose workers to errors. Therefore, it would be beneficial if the clients proposed projects with fewer risks. According to BP1-GW, their training must prepare them adequately to prevent accidents. They pointed out that while it is easier for the safety manager to instruct them on safe working practices, the reality of the situation is more complex than is represented during training. By incorporating learning-by-doing into safety training, the safety manager can provide examples of how to avoid making decisions that could put someone at risk. This is because learning-by-doing offers a realistic example of safety practice on the construction project.

Most of the multiple case studies interviewees reported that their employers had established safety management teams on their project sites entrusted with maintaining a safe working environment. The safety management team develops systems to identify hazards, improve safety procedures, and reduce accidents. This is a result of the consistently high accident rates in South Africa's construction industry, which are widely documented. As a result, a safety management team was established to develop an accident-prevention strategy. Some of the interviewees were quoted as follows:

> *It is the responsibility of contractors to promote a safe working environment, and the safety manager is responsible to teach the workers to know how to work safely and to understand the impact of the identified hazards and how to prevent such hazards on their project sites (BP2-SM).*
>
> *Learning about safety would help our workers to understand the importance of safety compliance and that safe practice is not a responsibility of one individual but the responsibility of everyone working on a project site (BP3-SM).*
>
> *The construction industry is very complex and dangerous, and it is important that our people working in construction to learn how to work safely and also learn how to operate machines, which often results in injuries. This is why we must have a safety manager to promote a safety culture on our sites (BP1-CM).*

The majority of those who participated in the interviews agreed that incorporating learning-by-doing on site would reduce human error and improve the safety system. In Table 4.3, the factors that influence learning-by-doing are divided into two categories: workers and management induced factors.

Table 4.3 Learning-by-doing induced factors

Human error-induced factors	Learning induced factors	Sources
Workers-induced error factor	Promoting safety training or learning on sites;	Multiple case studies
	Improving the workers' working experience through skills transfer;	Multiple case studies
	Improving communication between site management and workers;	Multiple case studies
	Motivating the workers to work safely;	Multiple case studies
	Transferring the knowledge and skills of construction work (i.e., from artisans to general workers), and	Multiple case studies
	Improving the workers' attitudes towards safety	Multiple case studies
Management-induced error factor	Making safety training facilities available on sites;	Multiple case studies
	Adopting modern technology to improve safety learning on sites;	Multiple case studies
	Having an appropriate SMS plan on sites;	Multiple case studies
	Providing different safety training courses for site management and general workers;	Multiple case studies
	Making safety training resources available on sites, and	Multiple case studies
	Improving the working environment/conditions.	Multiple case studies

4.5   Discussion

Table 4.4 shows how the MS and SD were used to rank the factors influencing learning-by-doing. The internal reliability of the factors is acceptable with a 0.996 Cronbach's α. The 15 learning-by-doing induced factors are categorised into worker-induced and management factors as identified from the multiple case studies. It is observed in Table 4.4 that the MS of eight learning-induced factors was above the midpoint of 3.00 in the worker-induced error factor. This result indicates that addressing the identified learning-by-doing induced factors would help to reduce human errors associated with the workers in construction. For example, workers' attitudes toward safety training and learning are ranked the highest with an MS of 4.07 and SD of 1.206, while a rewards system for learning outcomes is ranked the lowest with an MS of 3.12 and SD of 1.429. Among the learning-induced factors in this category were the knowledge and skills of workers, poor communication between workers and the supervisors, employers' motivation, personal working experience, commitment to learning about safety, and workers' self-confidence.

The MSs of seven management-induced error factors were above the midpoint of 3.00. This result indicates that addressing the seven learning-induced factors would help reduce construction errors. For example, management roles and

Table 4.4 Factors influencing how learning-by-doing could reduce human errors

Learning-by-doing induced workers' error factors	MS	SD	Rank
Workers' attitudes toward safety training/learning	4.07	1.206	1
Knowledge and skills of workers	3.91	1.235	2
Poor communication between workers and supervisors	3.89	1.340	3
Employers' motivation	3.73	1.285	4
Personal working experience	3.70	1.309	5
Safety learning commitment	3.58	1.298	6
Workers' self-confidence	3.45	1.329	7
Rewards system for learning outcomes	3.12	1.429	8
Learning-by-doing induced management error factors	MS	SD	Rank
Management roles and responsibilities in safety training/ learning	3.78	1.250	1
The working environment/conditions	3.67	1.213	2
Availability of safety training/learning resources	3.61	1.325	3
Standard of training facilitators (safety manager, foreman, construction manager, and site engineer)	3.59	1.297	4
Safety training/learning opportunities	3.48	1.334	5
Availability of modern technology to improve safety learning	3.16	1.471	6
Standard of training facilities on sites	3.06	1.473	7
Cronbach's α	**0.996**		

responsibilities in safety training and learning were ranked the highest with an MS of 3.78 and SD of 1.250. In contrast, the standard of training facilities on sites is ranked the lowest at an MS of 3.06 and SD of 1473. Among the learning-induced factors in this category were the working environment and conditions, the availability of safety training and learning resources, the standard of training facilitators (safety manager, foreman, construction manager, and site engineer), safety training and learning opportunities, and the availability of modern technology to improve safety learning.

Both the textual data in section 4.1 and the statistical data in this section suggest that integrating learning-by-doing into the safety management system would help reduce human error. Learning-by-doing could convey job knowledge and abilities in an informal training program. For instance, through observation, practice, and mentorship, people in construction can be taught the fundamentals of working safely. Safety managers or people in construction should be aware that learning-by-doing could reduce the likelihood of a hazard becoming a risk on their site. According to interviewees, safety training is provided at the beginning of the project. By combining it with learning-by-doing, the workers will learn about the safety requirements as they carry out their tasks. Therefore, learning-by-doing on a construction site should provide people in construction with opportunities for hands-on training to reduce human error.

4.6 Chapter summary

This chapter presents learning-by-doing to reduce human errors on construction sites. The data from the multiple case studies show that human error is an issue in construction projects and includes both s workers and management. According to several studies, the human error resulting from accidents is caused by actions, perceptions, and decision-making (cognitive biases or heuristics; Reason, 2000). Thus, multiple case studies have indicated that human error is common in the workplace and must be addressed. The chapter proposes using learning-by-doing to minimise the likelihood of human error in construction operations.

References

Ahmad R, Masse C, Jituri S, et al. (2018) Alberta Learning Factory for training reconfigurable assembly process value stream mapping. *Procedia Manufacturing* 23: 237–242. Available from: https://doi.org/10.1016/j.promfg.2018.04.023

Bruce BC and Bloch N (2012) Learning by doing. In: Seel SM (Ed.), *Encyclopedia of the sciences of learning*. Boston: Springer. Available from: https://doi.org/10.1007/978-1-4419-1428-6_544

Cabello AT, Martínez-Rojas M, Carrillo-Castrillo JA, et al. (2021) Occupational accident analysis according to professionals of different construction phases using association rules. *Safety Science* 144: 105457. Available from: https://doi.org/10.1016/j.ssci.2021.105457

Center for Construction Research and Training (CPWR) (2022, May) *Fatal and nonfatal injuries in construction*. Available from: www.cpwr.com/research/data-center/data-dashboards/fatal-and-nonfatal-injuries-in-construction/

Doytchev D and Hibberd R E (2009) Organizational learning and safety in design: Experiences from German industry. *Journal of Risk Research* 12(3–4): 295–312. Available from: https://doi.org/10.1080/13669870802604307

Fang W, Luo H, Xu S, et al. (2020) Automated text classification of near-misses from safety reports: An improved deep learning approach. *Advanced Engineering Informatics* 44: 101060. Available from: https://doi.org/10.1016/j.aei.2020.101060

Fang W, Wu D, Love PE, et al. (2022) Physiological computing for occupational health and safety in construction: Review, challenges and implications for future research. *Advanced Engineering Informatics* 54: 101729. Available from: https://doi.org/10.1016/j.aei.2022.101729

Federated Employers Mutual Assurance Company (FEM) (2022, April 26) *FEM registers alarming number of injuries in local construction industry*. Available from: https://zeroisnoaccident.co.za/fem-registers-alarming-number-of-injuries-in-local-construction-industry/

Hancock DR and Algozzine B (2011) *Doing case study research: A practical guide for beginning researcher*. New York: Teachers College Press.

Horvath D, Klamar A, Keith N, et al. (2021) Are all errors created equal? Testing the effect of error characteristics on learning from errors in three countries. *European Journal of Work and Organizational Psychology* 30(1): 110–124. Available from: https://doi.org/10.1080/1359432X.2020.1839420

Jonathan M, Bussier P and Chong H-Y (2022) Relationship between safety measures and human error in the construction industry: Working at heights. *International Journal of Occupational Safety and Ergonomics* 28(1): 162–173. Available from: https://doi.org/10.1080/10803548.2020.1760559

Li X, Guo Y, Ge F-L, et al. (2023) Human reliability assessment on building construction work at height: The case of scaffolding work. *Safety Science* 159: 106021. Available from: https://doi.org/10.1016/j.ssci.2022.106021

Love PE (2021) Creating a mindfulness to learn from errors: Enablers of rework containment and reduction in construction. *Developments in the Built Environment* 1: 100001. Available from: https://doi.org/10.1016/j.dibe.2019.100001

Mohd NI, Ali KN, Bandi S, et al. (2019) Exploring gamification approach in hazard identification training for Malaysian construction industry. *International Journal of Built Environment and Sustainability* 6(1): 51–57. Available from: https://doi.org/10.11113/ijbes.v6.n1.333

Plano Clark V and Ivankova NV (2016) *Mixed method research: A guide to the field*. Los Angeles, CA: Sage Publishing.

Rafieyan A, Sarvari H, Beer M, et al. (2022) Determining the effective factors leading to incidence of human error accidents in industrial parks construction projects: Results of a fuzzy Delphi survey. *International Journal of Construction Management*. Available from: https://doi.org/10.1080/15623599.2022.2159630

Reason J (2000) Human error: Models and management. *British Medical Journal* 320: 768–770.

Reason J (2016) *Organizational accidents revisited*. New York: CRC Press.

Reese H (2011) The learning-by-doing principle. *Behavioral Development Bulletin* 11: 1–19.

Xu S, Ni QQ, Zhang M, et al. (2019) A personalized safety training system for construction workers. In: *Proceedings of the international conference on smart infrastructure and construction 2019 (ICSIC): Driving data-informed decision-making* (pp. 321–326). Available from: https://doi.org/10.1680/icsic.64669.321

Ye G, Wang Y, Zhang Y, et al. (2019) Impact of migrant workers on total factor productivity in Chinese construction industry. *Sustainability* 11(926). Available from: https://doi.og/10.3390/su11030926

5 Managing health and safety on construction sites

5.1 Introduction

The construction industry is increasingly under client pressure to provide projects that adhere to occupational health and safety (OHS) regulations. Clients typically make important decisions about project budgets, schedules, objectives, and performance standards that could impact construction projects' OHS (Lingard et al., 2019). Clients play a major role in improving OHS performance in construction. Because of this, it is no longer up to contractors to decide whether to implement OHS legislation and related regulations. For example, in South Africa, clients demand that contractors adhere to the Construction Regulations 2014 (Republic of South Africa. Department of Labour, 2014). The 2014 version of the Construction Regulations aim to guarantee that the people involved in any construction project, regardless of size, scope, or whether it is being modified or constructed for the first time, are protected from an OHS problem (Kilian, 2014).

Over a decade ago, Smallwood and Haupt (2007) highlighted that the Construction Regulations define the duties of clients, designers, and contractors. The same view holds sway today as the 2014 Construction Regulations (Republic of South Africa. Department of Labour, 2014) contend that clients are responsible for formulating, documenting, and implementing project-specific health and safety (H&S) specifications based on the baseline risk assessment tool. Additionally, clients should stop work if it puts people in danger or if work violates H&S requirements (Department of Labour, 2014). This results from the focus on H&S placed throughout construction projects' design and planning phases (Lingard & Rowlinson, 2005).

Notably, accidents still often occur in construction operations despite legislative interventions. According to Hlati (2019), the Federated Employers Mutual (FEM) in South Africa estimated that in 2018, out of a total of 83,384 accidents in construction, 65 resulted in fatalities. Some accidents are caused by inadequate supervision and management (Department of Labour, 2017). The risky nature of operations and the poor working conditions result in fatalities and injuries (Jabbari & Ghorbani, 2016).

To address the stated OHS issue, this chapter presents a study that focused on the impact of safety managers on the OHS on construction projects. The focus

DOI: 10.1201/9781003365341-5

is premised on the notion that safety managers are responsible for establishing preventive and corrective H&S measures on construction sites (Reese & Eidson, 2006). The safety manager also designs an effective safety management system (SMS) to manage hazards and risks on construction sites. The same site management team member must effectively communicate H&S-related matters to construction workers (Tappura et al., 2017). Creating an efficient SMS includes preventing workers from acting in a way that could result in accidents, ensuring that issues are identified and addressed, and ensuring that occurrences are recorded and handled properly by the safety manager (Aksorn & Hadikusumo, 2008).

5.2   Safety managers and their function on construction sites

The construction industry contributes significantly to the global economy, employs around 7% of the working population globally, and produces about 6% of the gross domestic product (GDP; Babalola et al., 2023). Despite having a large impact on jobs and the global economy, the industry is one of the most hazardous owing to inherent hazards. Construction accidents frequently result in serious injury, death, and major financial loss (Rafindadi et al., 2023). The high injury rates imply that significant accident compensation, pain, loss, and suffering occur in construction. There has been much research into the underlying causes and contributing factors of construction site accidents and unsafe working conditions. Some causes cited in the literature include ignorance, a lack of safety training, human error, untrained personnel, inadequate supervision, negligence, apathy, outright recklessness, and poor and ineffective management of sites (Abukhashabah et al., 2020). As a result, owing to ineffective work procedures, badly administered regulations, and difficult clients, several OHS programs were established to improve their practice and performance (Awwad et al., 2016). Safety planning at the commencement of construction projects, for instance, would help to identify hazards before work begins (Zhang et al., 2013). Safety planning would further engender a process of selecting appropriate safety measures and recognising all potential hazards at once (Zhang, 2019).

Moreover, a company's capacity to compete in the construction industry is enhanced by safety performance, while poor performance harms reputation and competitiveness (Gao et al., 2018). The cornerstone to a successful safety program in construction, according to Choudhry et al. (2008), is managerial commitment and involvement. The management of construction projects must prioritise safety. Although implementing safety rules and procedures may raise workers' proficiency in executing their work, it cannot ensure that they will do it safely, as they may need more awareness training (Mohammadi et al., 2018). Therefore, the way to manage OHS is through training. OHS training is effective in creating and maintaining awareness. It includes training because employees must be more aware of potential risks (Zhu et al., 2022). The industry needs effective training procedures to improve construction workers' awareness of OHS (Demirkesen & Arditi, 2015). For example, studying first aid appears to reduce workers' preparedness to comply with construction site safety risk levels while enhancing their knowledge that unsafe behaviour among workers is a crucial component of preventing accidents. Additionally, workers who have received first aid training will better accept personal responsibility for safety to prevent accidents (Lingard, 2002). This is due to the possibility that first aid training will improve the attitudes of construction workers toward H&S procedures (Lingard, 2017).

Every major construction project should establish an SMS to prevent accidents (Choudhry et al., 2008). Having an SMS means having the right personnel to deploy it on sites. Such personnel will focus on removing inadequate safety management elements that result in fatalities and incur accident-related costs (Lai et al., 2011). Having an SMS means employing a safety manager to drive its effectiveness on a construction site. The safety manager would facilitate several potentially life-saving discussions through SMS to encourage a safe and healthy working environment (Choudhry et al., 2008). The safety manager would also emphasise that successful H&S records and practices foster a happy, risk-free, and effective work environment on construction sites.

According to Li et al. (2015), most accidents on construction projects are caused by unsafe behaviour, which safety managers must discuss directly with the workers. As a result, to reduce the likelihood of accidents, safety managers must consider several H&S management techniques (Gunter, 2007). For example, clients, employees, and the safety manager should all work together when preparing safety measures. The safety manager also needs to provide models of the proper safety procedures that should be implemented (Gunter, 2007).

5.3 Research methods

This study investigated the impact of safety managers on the OHS of construction projects. The data were collected in Bloemfontein, South Africa, using semi-structured and focus group interviews through the case study research approach (Yin, 2018). A case study approach was used to investigate how safety managers impacted the OHS on construction projects. This is because case studies focus on the problem (Yin, 2018). The study investigated various situations to identify the differences and similarities concerning the impact of safety managers on the OHS on construction projects. Three building projects formed the cases (Table 5.1). Building projects were selected because they expose workers to various hazards, and the safety manager oversees H&S on these projects (Zhang, 2019).

The interviewees' backgrounds in the multiple case studies are depicted in Table 5.1. To find similarities between the issues encountered in each case study, semi-structured interviews were conducted with operatives working on the three building projects. Purposive sampling was used to select the interviewees (Maxwell, 2013). The instrument for data collection used open-ended questions. The same format was followed for the focus group sessions in the multiple case studies. The interviewees in each case study were informed of the study's objective so they could understand the phenomenon. The interviews were conducted during lunchtime and lasted between 30 and 45 minutes. Each interview session was

Table 5.1 Multiple case study project details

Case studies	Code	Interviewees	Code	Number
Building Project 1	BP1	Construction manager	BP1-CM	1
		Safety manager	BP1-SM	1
		Site engineer	BP1-SE	1
		Senior foreman	BP1-SF	1
		Junior foreman	BP1-JF	1
		Group of artisans	BP1-GA	4
		Group of workers	BP1-GW	9
Building Project 2	BP2	Safety manager	BP2-SM	1
		Construction manager	BP2-CM	1
		Senior site engineer	BP2-SSE	1
		Senior site quantity surveyor	BP2-SSQS	1
		Senior foreman	BP2-SF	1
		Junior site engineer	BP2-JSE	1
		Junior foreman	BP2-JF	1
		Student supervisor	BP2-SS	1
Building Project 3	BP3	Construction manager	BP3-CM	1
		Safety manager	BP3-SM	1
		Site engineer	BP3-SE	1
		Senior foreman	BP3-SF	1
		Junior foreman	BP3-JF	1
		Group of artisans	BP3-GA	7
Total interviewees				**39**

recorded on a cell phone and the recordings were then transcribed. The qualitative data were thematically analysed after 39 interviews in total had been conducted.

5.4   Results

The interviewees in the multiple case studies reported that their employers had hired a safety manager to oversee H&S on their construction projects. They stated that a safety manager was appointed to establish an SMS (known as safety files to them) that would identify potential hazards and prevent injury to managers, site workers, visitors, and the public. The safety manager would ideally promote a safe and healthy workplace by following H&S policies and regulations on construction projects.

5.4.1   Tasks of safety managers

A BP3-SSF added that in addition to establishing an SMS, the safety manager is tasked with developing a safety training program to raise workers' awareness of potential accident causes. The interviewees further indicated that safety managers should regularly engage in safety training to remind workers of their employers' H&S policies and regulations. The BP2-SM concurred with the BP3-SSF's assertion while emphasising that the safety manager develops training that aligns with the risk assessment tool. With the help of this training, the workers would be able to understand the effects of the risks identified and learn how to prevent accidents caused by those risks.

A BP2-SSQS contended that safety training provides workers with the information they need to prevent accidents on construction sites. Through safety training on construction sites, workers can learn how to carry out their duties in line with H&S regulations. If the safety manager neglects to provide efficient safety training in line with the established H&S regulations, accidents would be associated with human error, according to a BP3-SE. This is because workers can break the law inadvertently when performing tasks related to construction since they need to understand the H&S laws and regulations. According to the BP1-CM, safety training can establish a foundation for reducing human error. The safety manager would then be able to identify the workers' issues and address them with safety training.

A BP2-CM reported that modifying the workers' unsafe behaviour through safety training is essential because the drive to perform faster frequently caused people to work unsafely. Workers' unsafe behaviour is associated with errors in the construction industry because it includes disregarding the H&S standards and regulations required on construction projects. To achieve high levels of productivity, the supervisor constantly puts pressure on the BP1-GW to finish the job that has been assigned as soon as possible. However, a BP3-GA indicated that their safety training had helped them understand how to perform under pressure because their supervisors count on them to be highly productive. They expressed gratitude to their safety manager for always teaching them to follow H&S guidelines, regardless of the pressure they might be under to attain high productivity levels. Adopting a hazardous approach that could lead to accidents would not help them impress their supervisors with high production. Only trained personnel should operate machinery and other equipment, according to a BP1-SM. This was owing to the increased likelihood of an accident should some workers use equipment with which they needed to be more familiar. Such behaviour could cause errors that could result in accidents.

5.4.2   *Worker empowerment*

According to the BP1-SM, safety managers must understand how empowering and encouraging workers can reduce accidents on construction sites. This is because it is assumed that motivated workers will follow H&S regulations. The safety culture is related to worker motivation, according to the BP3-JFM and BP3-SS. They underlined that the safety toolbox helps workers understand the significance of H&S, which clarifies why everyone on their sites wears PPE and walks in designated walking areas while construction is underway. Despite safety measures, the BP1-CM emphasised that the safety manager must frequently monitor H&S on construction projects. As a result, safety monitoring could be combined with safety training to increase worker motivation and safety awareness.

Most of the interviewees in the multiple case studies outlined the impact of safety managers in OHS on construction projects. The factors that impact safety managers in OHS on construction projects include the following:

- Having an effective SMS plan;
- Having a daily safety toolbox talk;
- Motivating the workers to think positively and improving perceptions about safety;
- Having an appropriate safe work procedure and risk assessment plan;
- Creating a safe working environment through good housekeeping;
- Having regular safety training for the workers;
- Having regular safety site inspections; and
- Hiring highly skilled safety managers (instructors) on sites.

5.5   Discussion

This chapter has outlined how safety managers could influence the OHS of construction projects. More often, the influence starts with designing an appropriate SMS in compliance with the regulations for a project. This function of a safety manager is nearly universal as it is replicated in developed and developing countries (Reese & Eidson, 2006). The safety manager also promotes a safe workplace through effective preventative and corrective strategies. According to a BP3-SSF, the safety manager is also tasked with developing a safety training program that will assist people in construction to become aware of potential accident causes. According to Hlati (2019), FEM estimated that in 2018, the South African construction industry recorded 83,384 accidents. Thus, the safety manager needs to provide safety training to assist the workers in preventing accidents.

Therefore, the safety manager must develop an SMS promoting a safe and healthy workplace, including compliance with H&S policies and regulations on construction projects. When H&S policies and regulations are not followed, accidents happen. For this reason, safety managers must constantly monitor the effectiveness of SMS on construction projects. Abukhashabah et al. (2020) asserted that most construction-related accidents and injuries are caused by negligence, a lack of safety training, human error, untrained workers, inadequate supervision, neglect, apathy, and flagrant recklessness. The reported causes of accidents, including human error and a lack of safety training, point to insufficient SMS implementation.

According to Reese and Eidson (2006), safety managers should implement effective and practical preventive and corrective methods for promoting H&S and a safe workplace. Figure 5.1 outlines a framework for assessing the impact of the safety manager on monitoring OHS on construction projects. It should be noted that the first step is for the contractor to appoint a competent safety manager to establish an SMS. The safety manager should understand the OHS policies and regulations required for construction projects. Thereafter, the safety manager should design an effective SMS, known as safety files, to be implemented in construction projects. The effective SMS should comprise hazard identification, a risk assessment plan,

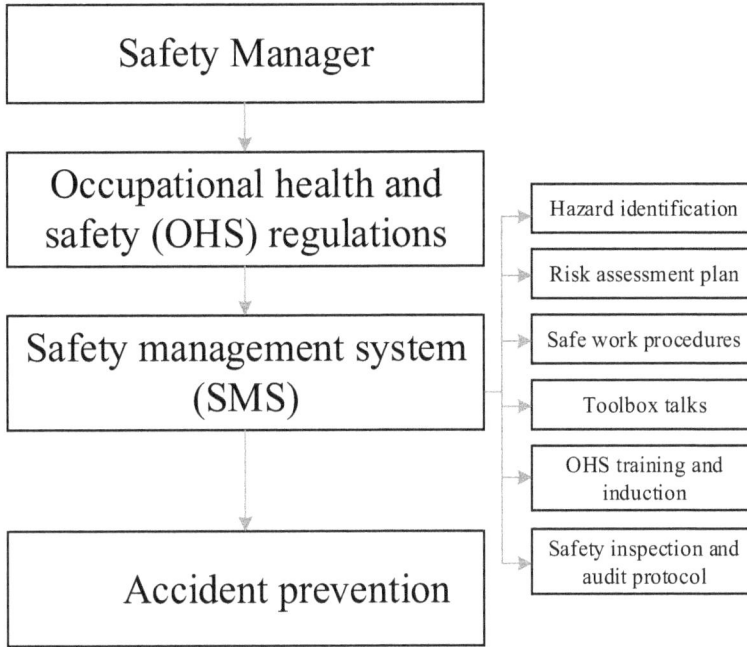

Figure 5.1 A safety manager illustrated work effect on construction projects

safe work procedure, toolbox talk, OHS training and induction, safety inspection, and audit protocols. As a result of an effective SMS, the safety manager would then have the ability to monitor OHS and prevent accidents on construction projects.

5.6   Chapter summary

The chapter postulates that safety managers influence the OHS on construction projects. This is because safety managers must establish the SMS, which would assist in monitoring the OHS on construction projects. Figure 5.1 shows how to monitor the safety managers' influence on monitoring the OHS to prevent accidents on construction projects. Contractors must pay special attention to how safety managers deal with OHS on construction projects. This would help in monitoring the OHS and preventing errors from causing accidents.

References

Abukhashabah E, Summan A and Balkhyour M (2020) Occupational accidents and injuries in construction industry in Jeddah City. *Saudi Journal of Biological Sciences*: 1993–1998. Available from: https://doi.org/10.1016/j.sjbs.2020.06.033

Aksorn T and Hadikusumo BH (2008) Critical success factors influencing safety program performance in Thai construction projects. *Safety Science* 46(4): 709–727.

Awwad R, Souki OE and Jabbour M (2016) Construction safety practices and challenges in a Middle Eastern developing country. *Safety Science* 83: 1–11.

Babalola A, Manu P, Cheung C, et al. (2023) A systematic review of the application of immersive technologies for safety and health management in the construction sector. *Journal of Safety Research*. Available from: https://doi.org/10.1016/j.jsr.2023.01.007

Choudhry RM, Fang D and Ahmed SM (2008) Safety management in construction: Best practices in Hong Kong. *Journal of Professional Issues in Engineering Education and Practice* 134(1): 20–32.

Demirkesen S and Arditi D (2015) Construction safety personnel's perceptions of safety training practices. *International Journal of Project Management* 33: 1160–1169.

Gao R, Chan A, Lyu S, et al. (2018) Investigating the difficulties of implementing safety practices in international construction projects. *Safety Science* 108: 39–47.

Gladwin C and Adam C (2014, November 20) *Impact of new OHS act construction regulations*. Retrieved from: www.bizcommunity.com/Article/196/360/121602.html

Gunter RE (2007) Checking safety in technology education. *The Technology Teacher* 66(6): 5–13.

Hlati O (2019, October 16) *Safety in construction industry under the spotlight*. Available from: www.iol.co.za/capetimes/news/safety-in-construction-industry-under-the-spotlight-35098202

Jabbari M and Ghorbani R (2016) Developing techniques for cause-responsibility analysis of occupational accidents. *Accident Analysis and Prevention* 96: 101–107.

Kilian A (2014, March 21) *New construction regulations stipulate more stringent safety obligations*. Available from: www.engineeringnews.co.za/print-version/changes-to-construction-regulations-of-the-occupational-health-and-safety-act-ohsa-2014-03-07

Lai DN, Liub M and Ling FY (2011) A comparative study on adopting human resource practices for safety management on construction projects in the United States and Singapore. *International Journal of Project Management* 29(8): 1018–1032.

Li H, Lu M, Hsu S, et al. (2015) Proactive behavior-based safety management for construction safety improvement. *Safety Science* 75: 107–117.

Lingard H (2002) The effect of first aid training on Australian construction workers' occupational health and safety motivation and risk control behavior. *Journal of Safety Research* 33(2): 209–230.

Lingard H (2017) First aid and occupational health and safety: The case for an integrated training approach. *Journal of the International Society for Burn Injuries*: 111–117.

Lingard H, Oswald D and Le T (2019) Embedding occupational health and safety in the procurement and management of infrastructure projects: Institutional logics at play in the context of new public management. *Construction Management and Economics* 37(10): 567–583. Available from: https://doi.org/10.1080/01446193.2018.1551617

Lingard H and Rowlinson S (2005) *Occupational health and safety in construction project management*. New York: Routledge.

Maxwell J (2013) *Qualitative research design: An interactive approach*, vol. 3. Los Angeles: Sage Publications.

Mohammadi A, Tavakolan M and Khosravi Y (2018) Factors influencing safety performance on construction projects: A review. *Safety Science* 19: 382–397.

Rafindadi AD, Shafiq N, Othman I, et al. (2023) Data mining of the essential causes of different types of fatal construction accidents. *Heliyon* 9(2): e13389. Available from: https://doi.org/10.1016/j.heliyon.2023.e13389

Reese CD and Eidson JV (2006) *Handbook of OSHA construction safety and health*, 2nd ed. New York: Taylor & Francis Group.

Republic of South Africa. Department of Labour (2014) *Occupational health and safety act, 1993: Construction regulations, 2014.* Pretoria, RSA: Government Gazette.

Republic of South Africa. Department of Labour (2017, March 9) *Labour on injuries and fatalities in SA construction sector.* Available from: www.gov.za/speeches/sa-construction-sector-9-mar-2017-0000

Smallwood JJ and Haupt TC (2007) Impact of the South African construction regulations on construction health and safety. *Journal of Engineering, Design and Technology* 5(1): 23–34.

Tappura S, Nenonen N and Kivistö-Rahnasto J (2017) Managers' viewpoint on factors influencing their commitment to safety: An empirical investigation in five finish industrial organisations. *Safety Science* 96: 52–61.

Yin R (2018) *Case study research and applications: Design and methods*, 6th ed. Los Angeles, CA: Sage Publications.

Zhang PL (2019) An agent-based modeling approach for understanding the effect of worker management interactions on construction workers' safety-related behaviors. *Automation in Construction* 97: 29–43.

Zhang S, Teizer J, Lee JK, et al. (2013) Building information modeling (BIM) and safety: Automatic safety checking of construction models and schedules. *Automation in Construction* 29: 183–195.

Zhu X, Li RY, Crabbe J and Sukpascharoen K (2022) Can a chatbot enhance hazard awareness in the construction industry? *Frontiers in Public Health* 10. Available from: https://doi.org/10.3389/fpubh.2022.993700

6 Minimising human errors on construction sites

6.1 Introduction

The construction industry continues to have one of the highest accident rates. According to the Health Safety and Executive (HSE; 2018), most accidents reported in the construction industry are caused by workers' unsafe behaviour. As a result, safety is a top priority in the construction industry, while workers' unsafe behaviour is one of the leading causes of construction accidents (Zhang et al., 2023). For example, more than 80% of accidents on construction sites are caused by workers' unsafe behaviour, including being hit by moving vehicles, becoming entangled in or between moving machinery, and falling from heights (Zhou et al., 2019).

Workers' unsafe behaviour on construction sites happens because of human error (Ahamed & Mariappan, 2023). According to Liu et al. (2020), human error is the main contributing factor to accidents. Errors occur when the human mind tries to organise an action, solve an issue, or carry it out incorrectly (Kandregula & Le, 2020). Falling from a height is one accident that regularly occurs in the construction industry and is caused by human error (Liu et al., 2019). The rate of human errors must be reduced if the safety management system (SMS) on construction sites is to be improved. Significant improvements have been made to the SMS implemented on construction sites to reduce damage severity and human error frequency (Liu et al., 2019).

A SMS comprises individuals who work to establish and maintain a safe and healthy workplace, as well as occupational health and safety (OHS) policies and regulations (Xu et al., 2023). SMSs are institutionalised with policies, goals, objectives, processes, responsibilities, and other safeguards to deal with known risks. They are categorised based on their components, such as toolbox talks, hazard identification, safety inspection, OHS training, safe work procedure, risk assessment plans, and safety reporting systems (Mollo et al., 2021). SMS focuses on minimising errors rather than addressing them (Pilanawithana et al., 2022). As a result, various internal and external factors, such as occurrences or acts other than human factors that increase the likelihood of an accident, may impact the SMS (Pereira et al., 2018). Consequently, it is necessary to anticipate risks to reduce human error effectively and improve the capabilities of an active SMS in construction (Liu et al., 2020).

DOI: 10.1201/9781003365341-6

As a result, this concluding chapter of the book presents an integrated framework for reducing human error on construction sites. The SMS is integrated with the training-within-industry (TWI) program to develop a conceptual model to minimise human error on construction sites. An integrated SMS and TWI would reduce human error on construction sites. In this case, TWI would equip the safety manager with the knowledge required to improve SMS monitoring, which would help prevent construction site accidents.

6.2   TWI method revisited

TWI is a form of leadership education that gives the supervisor the skills necessary to manage, instruct, and reinforce the duties of their subordinates (Allen, 1919). The United States Bureau of Training used it between 1940 and 1945 to help train contractors for the war effort (Bianchi & Giorcelli, 2021). TWI has existed since World War II; it is not a new way of thinking (Pascual, 2017). In industry, training aims to improve worker behaviour to promote efficiency and higher performance standards. Also, training ensures that individuals experience in a methodical way the attitude, knowledge, and behaviour pattern needed to do a job well and provide increased productivity (Renukappa et al., 2021). Training is a deliberate process to modify one's behaviour, knowledge, or attitude to excel at any activity or group of activities (Masadeh, 2012). Moreover, training at work aims to support individual growth as well as meet the organisation's labour demands. As such, receiving training results in immediate advantages, including improved problem-solving abilities, a more knowledgeable and effective team, easier access to suitable applicants, and improved employee relations (Renukappa et al., 2021).

After the introduction of factory schools, which introduced industrial conditions into the classroom around 1800, training procedures experienced a dramatic upheaval (Fenner et al., 2018). Employees could learn from the same trainer instead of on-the-job training in a classroom setting, which has both disadvantages and advantages, such as poor task memory and learning transfer. The vestibule training approach of the early 1900s combines the benefits of in-class instruction and practical experience (Fenner et al., 2018). With this approach, the training room would be situated close to the working environment and be equipped with the same production-related machinery. Hence, this approach had an impact on the development of the TWI program.

The TWI program offered voluntary and free in-plant management training to contractors engaged in war production between 1940 and 1945, supported by the US government. The TWI program was designed to be productive because it was a service to assist contractors involved in war production during WWII (Allen, 1919). The objective of the TWI program is to increase productivity and eliminate anything that impedes the flow of production in the workplace (Pascual, 2017). The lofty goal of the original TWI program was to provide management training to all US military contractors; however, this initial concept proved unworkable owing to a constrained budget and a manpower shortage. Because of the limited funding for the TWI program, it was decided to teach only businesses willing to participate in the training (Bianchi & Giorcelli, 2021).

The TWI program's development considerably increased the industrial output of military supplies (Singh & Ballerio, 2016). As a result, each training program created using TWI must adhere to the following principles:

- Training must be defined;
- Minimal technical presentation should be used during training;
- Learning-by-doing must be the foundation of all training; and
- The training must have a multiplier effect, enabling the learner to pass on the skill in exactly the same form as received (Pascual, 2017).

In addition, the development of the TWI program, according to Grip and Sauermann (2013), was based on the Three J-Program, which comprises Job Instruction (JI), Job Relations (JR), and Job Methods (JM). The three J-modules are described in detail below:

- Job Instruction (JI): The main goal of this training program was to teach participants (employers) how to complete tasks swiftly, correctly, and safely. As a result, trained companies began creating standard operating procedures, enhancing job safety procedures, improving lighting, keeping the factory floor clean to prevent accidents and facilitate the movement of materials, maintaining machines regularly, and keeping track of the causes of malfunctions (Bianchi & Giorcelli, 2021).
- Job Relations (JR): This training program aimed to develop competent supervisors by emphasising collaboration. One approach was to teach supervisors how to build and maintain positive relationships with each person in the workplace (Singh & Ballerio, 2016). JR would, therefore, assist in resolving conflicts and issues that could emerge amongst team members or irate clients by taking the following four steps: gathering information, weighing options, deciding, acting, and evaluating outcomes (Pascual, 2017).
- Job Methods (JM): This training program enabled supervisors to evaluate and improve their working procedures. JM helped companies to continue improving their operating processes as a result. They started managing their inventories, production planning, and production tracking more skilfully to prioritise customer orders depending on delivery deadlines (Bianchi & Giorcelli, 2021).

6.2.1   *Conceptual model for reducing human error*

The conceptual model for minimising human error on construction sites is shown in Figure 6.1. The conceptual model assumes that human errors are mostly to blame for accidents (Liu et al., 2020). As a result, an ineffective SMS utilised on construction sites highlights human error. Therefore, improved SMS is necessary to minimise human error on construction sites (Liu et al., 2019). The TWI job

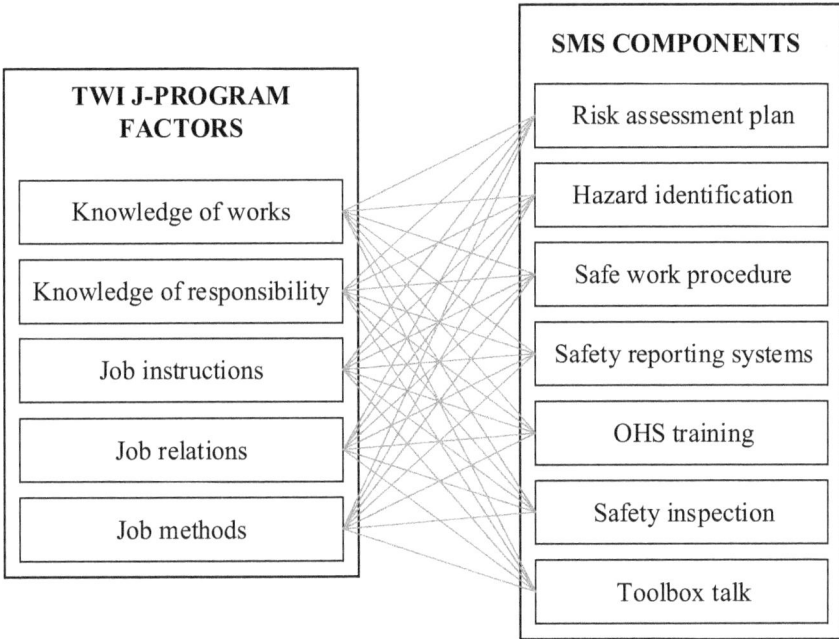

Figure 6.1 A conceptual model for minimising human error on construction sites

program factors and the SMS components standardised the conceptual model. for example, the safety manager will be facilitated in developing an SMS by merging the TWI job program and the SMS components. As a result, by integrating the TWI job program and the SMS components, the safety manager will be assisted in monitoring the SMS.

The TWI job program factors are combined with the SMS components to improve skills and knowledge in monitoring the SMS to minimise human error on construction sites. This is because the TWI program prepares supervisors with the abilities needed to coordinate, instruct, and enforce the responsibilities of their subordinates (Allen, 1919). The adoption of the TWI job program is influenced by the following:

- Knowledge of the work;
- Knowledge of responsibility;
- Job instruction;
- Job methods; and
- Job relations (Huntzinger, 2016).

The TWI job program factors and SMS components are interconnected to monitor each SMS component more effectively. This aligns with the idea that SMS minimises errors rather than addresses them (Pilanawithana et al., 2022), and that integrating the TWI job program will help mitigate this issue. Hence, enhanced SMS component monitoring could help minimise human error on construction sites. The SMS's components include the following:

- Toolbox talks;
- Hazard identification;
- Safety inspection;
- OHS training;
- Safe work procedures;
- Risk assessment plans; and
- Safety reporting systems (Mollo et al., 2021).

Each TWI job program factor connects to the SMS components, providing the safety manager with the skills and knowledge to improve the SMS on construction sites. For instance, job instruction will give the safety manager knowledge of the job that will assist in minimising human error. The safety manager's initiatives to promote respect for people on construction sites will result in better job relations. Also, knowledge of the job methods will assist the safety manager in continually improving the established SMS to minimise human error and prevent accidents.

6.3   Research methods

This research evaluated a conceptual model for minimising human error on construction sites. The conceptual model evolved through a quantitative research approach (Creswell & Clark, 2018). A research survey conducted among construction health and safety specialists in South Africa was used to obtain the quantitative data. Online questionnaires with closed-ended questions were used to produce the survey questionnaire. The link to the survey questionnaires was then sent to the Association of Construction Health and Safety Management (ACHASM) for distribution to its members.

ACHASM was created to serve as a representative and advisory body for everyone involved in the construction health and safety (CHS) business and is expressed in its definition as a legally registered non-profit organisation. CHS management specialists comprised the study's sample (Table 6.1). As they were all ACHASM members and had access to the survey link through the organisation's secretariat, it can be said that a simple random sample approach was used to select the survey respondents.

The survey questionnaire was designed as Creswell and Clark (2018) recommended. The survey questionnaires used a Likert scale of 0– 5, where 0 = Unreliable, 1 = A minor, 2 = Above minor, 3 = Neutral, 4 = Below major, and 5 = Significant. The quantitative data were analysed using the Statistical Program for the Social Sciences (SPSS). Descriptive statistical analysis uses mean scores (MSs) and standard deviations (SDs). The reliability of the data was tested using

Table 6.1 Research sample

Demographic profile	Category	Number (N)	Percentage (%)
Types of occupation	Construction health and safety agent	9	31.0
	Construction health and safety manager	12	41.4
	Construction health and safety officer	8	27.6
Total number of participants		**29**	**100.00**
Types of educational qualification	Matric	1	3.5
	Certificates	8	27.6
	National diploma	10	34.5
	Bachelor's degrees	5	17.2
	Honours degrees or PG Dip	3	10.3
	Master's degrees	2	6.9
Total number		**29**	**100.00**
Years of experience	1–5	3	10.3
	5–10	4	13.8
	10–15	12	41.4
	15–20	6	20.7
	20 or above	4	13.8
Total		**29**	**100.0**

Cronbach's α, as Creswell and Clark (2018) recommended. The survey questionnaires used a cross-sectional study design because it was carried out between 12 February and 26 April 2020. The total responses received from the online survey platform numbered 29, as shown in Table 6.1. In addition, the respondents' educational backgrounds and years of experience are indicated in Table 6.1.

6.4   Results

The evaluation of a conceptual model for minimising human error on construction sites is presented in this section. The conceptual model integrates the five TWI job program factors with the seven SMS components. The following investigations were used to assess a conceptual model:

- What the impact of knowledge of work related to the SMS components is on construction sites;
- What the impact of knowledge of responsibility related to the SMS components is on construction sites;
- How job instructions influence the SMS components on construction sites;
- How job methods influence the SMS components on construction sites; and
- How job relations influence the SMS components on construction sites.

6.4.1 Knowledge of work and SMS components on construction sites

The extent to which knowledge of work was related to the seven SMS components on construction sites is shown in Table 6.2. The statistical data, based on a Likert scale of 1–5, were analysed using SPSS. The MS of the integrated seven knowledge of work aspects was above the midpoint of 3.00, which indicated that the respondents perceived that knowledge of work would help the safety manager with a deep understanding of monitoring the SMS components on construction sites. The aspects evaluated included the impact of knowledge of work on safe work procedures with an MS of 3.93, the impact of knowledge of work on hazard identification with an MS of 3.89, the impact of knowledge of work on safety inspection with an MS of 3.18, the impact of knowledge of work on risk assessment plan with an MS of 3.81, the impact of knowledge of work on OHS training with an MS of 3.78, the impact of knowledge of work on toolbox talks with an MS of 3.52, and the of the impact of knowledge of work on safety reporting system with an MS of 3.41.

The results shown in Table 6.2 indicated that the data used to measure the extent to which knowledge of work was integrated to the SMS components were reliable (Cronbach's α =0.988).

6.4.2 Knowledge of responsibility and SMS components on construction sites

The extent to which knowledge of responsibility was related to the SMS components on construction sites is shown in Table 6.3. The statistical data, based on a Likert scale of 1–5, were analysed using SPSS. The MS of the seven knowledge of responsibility aspects was above the midpoint of 3.00, which indicated that the respondents perceived that knowledge of responsibility would be improved by the safety manager's coordinating SMS components on construction sites. The aspects evaluated included the impact of knowledge of responsibility on toolbox talk with an MS of 3.89, the impact of knowledge of responsibility on OHS training with an MS of 3.85, the impact of knowledge of responsibility on risk assessment plan with an MS of 3.81, the impact of

Table 6.2 Knowledge of work impact on the SMS components

Aspect	MS	SD	Rank
The impact of knowledge of work on safe work procedures	3.93	1.174	1
The impact of knowledge of work on hazard identification	3.89	1.281	2
The impact of knowledge of work on safety inspection	3.18	1.145	3
The impact of knowledge of work on a risk assessment plan	3.81	1.210	4
The impact of knowledge of work on OHS training	3.78	1.155	5
The impact of knowledge of work on toolbox talk	3.52	1.312	6
The impact of knowledge of work on safety reporting systems	3.41	1.118	7
Cronbach's α	**0.988**		

Table 6.3 Knowledge of responsibility impact on SMS components

Aspect	MS	SD	Rank
The impact of knowledge of responsibility on toolbox talk	3.89	1.219	1
The impact of knowledge of responsibility on OHS training	3.85	1.486	2
The impact of knowledge of responsibility on a risk assessment plan	3.81	1.178	3
The impact of knowledge of responsibility on hazards identification	3.74	1.318	4
The impact of knowledge of responsibility on safe work procedure	3.70	1.265	5
The impact of knowledge of responsibility on safety inspection	3.67	1.240	6
The impact of knowledge of responsibility on a safety reporting system	3.37	1.149	7
Cronbach's α	**0.985**		

knowledge of responsibility on hazards identification with an MS of 3.74, the impact of knowledge of responsibility on safe work procedure with an MS of 3.70, the impact of knowledge of responsibility on safety inspection with an MS of 3.67, and the impact of knowledge of responsibility on the safety reporting system with an MS of 3.37.

The results shown in Table 6.3 indicated that the data used to measure the extent to which knowledge of responsibility was integrated into the SMS components were reliable (Cronbach's α = 0.985).

6.4.3 Job instructions and SMS components on construction sites

The extent to which job instructions were related to the SMS components on construction sites is shown in Table 6.4. The statistical data, based on a Likert scale of 1–5, were analysed using SPSS. The MS of the integrated job instruction aspects was above the midpoint of 3.00, which indicated that the respondents perceived

Table 6.4 Job instructions impact on the SMS components

Aspect	MS	SD	Rank
The impact of job instructions on risk assessment plan	4.00	1.020	1
The impact of job instructions on safety training and education	3.85	0.784	2
The impact of job instructions on the hazard identification	3.81	1.021	3
The impact of job instruction on safe work procedures	3.81	1.021	4
The impact of job instructions on safety inspection	3.73	0.874	5
The impact of job instructions on toolbox talk	3.69	1.123	6
The impact of job instructions on safety reporting systems	3.54	1.102	7
Cronbach's α	**0.985**		

that job instructions would continue to help the safety manager have a deep understanding of implementing the SMS components on construction sites. The aspects evaluated included the impact of job instructions on a risk assessment plan with an MS of 4.00, the impact of job instructions on OHS training with an MS of 3.85, the impact of job instructions on the hazards identification with an MS of 3.81, the impact of job instructions on safe work procedures with an MS of 3.81, the impact of job instructions on safety inspection with an MS of 3.73, the impact of job instructions on toolbox talk with an MS of 3.69, and the impact of job instructions on safety reporting systems with an MS of 3.54.

The results shown in Table 6.4 indicated that the data used to measure the extent to which job instructions were integrated into the SMS components were reliable (Cronbach's α = 0.985).

6.4.4   *Job methods and SMS components on construction sites*

The extent to which job methods were related to the seven SMS components on construction sites is shown in Table 6.5. The statistical data, based on a Likert scale of 1–5, were analysed using SPSS. The MS of the integrated job method aspects was above the midpoint of 3.00, which indicated that the respondents perceived that job methods could help the safety manager improve the incorporation of the SMS components on construction sites. The aspects evaluated included the impact of job methods on a risk assessment plan with an MS of 4.04, the impact of job methods on hazard identification with an MS of 3.88, the impact of job methods on OHS training with an MS of 3.85, the impact of job methods on safety inspection with an MS of 3.62, the impact of job methods on safe work procedure with an MS of 3.5, the impact of job methods on safety reporting systems with an MS of 3.50, and the impact of job methods on toolbox talks with an MS of 3.50.

The results in Table 6.5 indicated that the data used to measure how well job instructions were integrated into the SMS components were reliable (Cronbach's α = 0.980).

Table 6.5 Job methods' impact on the SMS components

Aspect	MS	SD	Rank
The impact of job methods on risk assessment plan	4.04	0.916	1
The impact of job methods on hazard identification	3.88	1.143	2
The impact of job methods on OHS training	3.85	0.784	3
The impact of job methods on safety inspection	3.62	1.098	4
The impact of job methods on safe work procedure	3.58	1.064	5
The impact of job methods on safety reporting systems	3.50	1.105	6
The impact of job methods on toolbox talk	3.50	1.241	7
Cronbach's α	**0.980**		

Table 6.6 Job relations' impact on the SMS components

Aspect	MS	SD	Rank
The impact of job relations on toolbox talk	4.04	0.916	1
The impact of job relations on a risk assessment plan	4.04	1.038	2
The impact of job relations on hazard identification	3.96	1.076	3
The impact of job relations on OHS training	3.81	0.895	4
The impact of job relations on safe work procedures	3.77	1.070	5
The impact of job relations on safety inspections	3.65	1.093	6
The impact of job relations on a safety reporting system	3.54	0.989	7
Cronbach's α	**0.983**		

6.4.5   *Job relations and SMS components on construction sites*

The extent to which job relations were related to the seven SMS components on construction sites is shown in Table 6.6. The statistical data, based on a Likert scale of 1–5, were analysed by using SPSS. The MS of the integrated job relation aspects was above the midpoint of 3.00, which indicated that the respondents perceived that job relations would help the safety manager foster respect among workers by incorporating the SMS components on construction sites. The aspects evaluated included the impact of job relations on toolbox talk with an MS of 4.04, the impact of job relations on the risk assessment plan with an MS of 4.04, the impact of job relations on the hazard identification with an MS of 3.96, the impact job relations on OHS training with an MS of 3.81, the impact of job relations on safe work procedures with an MS of 3.77, the impact of job relations on safety inspections with an MS of 3.65, and the impact of job relations on safety reporting systems with an MS of 3.54.

The results shown in Table 6.6 indicated that the data used to measure the extent to which job relations were integrated into SMS components were reliable (Cronbach's α = 0.983).

6.5   Discussion

An enhanced version of the human error reduction model on construction sites is presented in this section based on the findings from the previous section. This conceptual model is important because human error is the main contributor to accidents (Liu et al., 2020). In addition, the unsafe behaviour among workers in construction also contributes to human error (Ahamed & Mariappan, 2023). To address human error causing accidents, the TWI job program factors are integrated with the SMS components. TWI is designed to eliminate anything hindering work (Pascual, 2017). After that, the TWI program would help the safety manager mitigate safety issues that can cause unsafe behaviour and human error.

Therefore, the integration of the TWI job program factors with the SMS components contributed to the updated human error reduction model on construction sites, as shown in Figure 6.2. For example, the conceptual model combines TWI job program factors and SMS components to create new knowledge, specifically

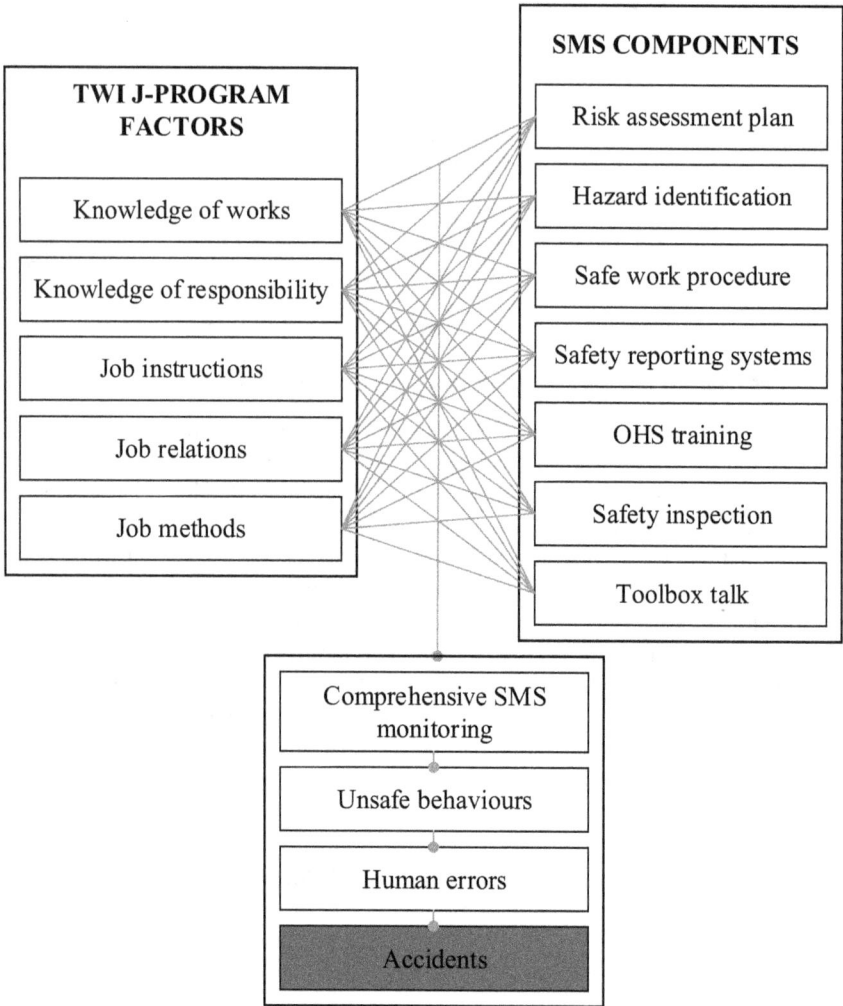

Figure 6.2 An enhanced conceptual model for minimising human errors on construction
sites

the interpretation and abilities of a safety manager to monitor the SMS components. Through comprehensive SMS monitoring, the safety manager could identify
unsafe behaviours among workers and minimise human error. If human error is
minimised, accidents can be prevented. The idea is consistent with efforts to reduce
the severity of damage and the potential for human error (Liu et al., 2019).

6.6   Chapter summary

The concluding chapter of this book describes how to reduce human error using
a model in which TWI and SMS are embedded. The overall intent is to reduce or

eliminate accidents on construction sites. A comprehensive SMS monitoring could assist in such an endeavour when it is connected to TWI job program factors. The human error reduction model could be beneficial in the following ways:

- Acknowledging that human error is a problem in construction operations;
- Identifying the TWI job factors that should be merged with the SMS components;
- Identifying the SMS components and integrating the TWI job program factors; and
- Providing guidelines for integrating the TWI job program factors with the SMS components to enhance SMS monitoring.

References

Ahamed N and Mariappan M (2023) A study to determine human-related errors at the level of top management, safety supervisors & workers during the implementation of safety practices in the construction industry. *Safety Science* 162: 106081. Available from: https://doi.org/10.1016/j.ssci.2023.106081

Allen E (1919) *The instructor, the man, and the job*. Philadelphia and London: JB Lippincott.

Bianchi N and Giorcelli M (2021) *The dynamics and spillovers of management interventions: Evidence from the training within industry program*. Cambridge: National Bureau of Economic Research.

Creswell JW and Clark VP (2018) *Designing and conducting mixed method research*, 3rd ed. Thousand Oaks, CA: Sage Publishing.

Fenner AE, Morque S, Sullivan J, et al. (2018) Emerging workforce training methods for the construction industry. In: *Proceedings of the construction research congress 2018* (pp. 608–618). New Orleans, LA: ASCE. Available from: https://doi.org/10.1061/9780784481271.059

Grip AD and Sauermann J (2013) The effect of training on productivity: The transfer of on-the-job training from the perspective of economics. *Educational Research Review* 8: 28–36.

Health and Safety Executive (HSE) (2018) *Improving health and safety outcomes in construction*. London: Health and Safety Executive.

Huntzinger J (2016) *The roots of lean: Training within industry: The origin of Japanese management and kaizen and other insights*. Indianapolis, IN: Lean Frontiers.

Kandregula SK and Le T (2020) Investigating the human errors in 4D BIM construction scheduling. In: *Proceedings of the construction research congress 2020: Project management and controls, materials, and contracts* (pp. 750–757).

Liu M, Chong HY, Liao PC, et al. (2019) Probabilistic-based cascading failure approach to assessing workplace hazards affecting human error. *Journal of Management in Engineering* 35(3): 04019006. Available from: https://doi.org/10.1061/(ASCE)ME.1943-5479.0000690

Liu M, Tang P, Liao PC, et al. (2020) Propagation mechanics from workplace hazards to human errors with dissipative structure theory. *Safety Science* 126: 104661. Available from: https://doi.org/10.1016/j.ssci.2020.104661

Masadeh M (2012) Training, education, development and learning: What is the difference? *European Scientific Journal* 8(9): 62–68. Available from: https://doi.org/10.19044/esj.2012.v8n10p

Mollo LG, Emuze F and Smallwood JJ (2021) Using a safety management system to reduce errors and violations. *Proceedings of the Institution of Civil Engineers – Municipal Engineer* 174(3): 136–143. Available from: https://doi.org/10.1680/jmuen.20.00017

Pascual AA (2017) *Training within industry in the emergency department: Team development to improve patient care and alleviate staff burnout.* Irvine, CA: Brandman University.

Pereira E, Ahn S, Han S, et al. (2018) Identification and association of high-priority safety management system factors and accident precursors for proactive safety assessment and control. *Journal of Management in Engineering* 34(1): 04017041. Available from: https://doi.org/10.1061/(ASCE)ME.1943-5479.0000562

Pilanawithana NM, Feng Y, London K, et al. (2022) Developing resilience for safety management systems in building repair and maintenance: A conceptual model. *Safety Science* 152: 105768. Available from: https://doi.org/10.1016/j.ssci.2022.105768

Renukappa S, Suresh S and Alosaimi H (2021) Knowledge management-related training strategies in kingdom of Saudi Arabia construction industry: An empirical study. *International Journal of Construction Management* 13(1): 713–723. Available from: https://doi.org/10.1080/15623599.2019.1580002

Singh R and Ballerio N (2016) TWI (Training within industry). In: Baroncelli C and Ballerio N (Eds.), *WCOM (World Class Operations Management)* (pp. 227–244). Switzerland: Springer International Publishing. Available from: https://doi.org/10.1007/978-3-319-30105-1_20

Xu J, Cheung C, Manu P, et al. (2023) Implementing safety leading indicators in construction: Toward a proactive approach to safety management. *Safety Science* 157: 105929. Available from: https://doi.org/10.1016/j.ssci.2022.105929

Zhang Z, Guo H, Gao P, et al. (2023) Impact of owners' safety management behavior on construction workers' unsafe behavior. *Safety Science* 158: 105944. Available from: https://doi.org/10.1016/j.ssci.2022.105944

Zhou C, Chen R, Jiang S, Zhou Y, Ding L, Skibniewski MJ and Lin X (2019) Human dynamics in near-miss accidents resulting from unsafe behavior of construction workers. *Physica A* 530: 121495. Available from: https://doi.org/10.1016/j.physa.2019.121495

Index

Page numbers in *italics* indicate figures and page numbers in **bold** indicate tables.

For Product Safety Concerns and Information please contact our EU
representative GPSR@taylorandfrancis.com
Taylor & Francis Verlag GmbH, Kaufingerstraße 24, 80331 München, Germany

www.ingramcontent.com/pod-product-compliance
Lightning Source LLC
Chambersburg PA
CBHW060322220326
41598CB00027B/4401